# Combating Cyber Threat

# Combating Cyber Threat

*Editor*

## Lieutenant General PK Singh, PVSM, AVSM (Retd)

(Established 1870)

United Service Institution of India

New Delhi

Vij Books India Pvt Ltd

New Delhi (India)

*Published by*

**Vij Books India Pvt Ltd**
(Publishers, Distributors & Importers)
2/19, Ansari Road
Delhi – 110 002
Phones: 91-11-43596460, 91-11-47340674
Fax: 91-11-47340674
e-mail: vijbooks@rediffmail.com


Copyright © 2018, United Service Institution of India, New Delhi


ISBN: 978-93-86457-81-3 (Hardback)
ISBN: 978-93-86457-82-0 (ebook)

# Contents

*Preface*                                                                  vii

1. Norms in Cyberspace: United Nations Group of
   Government Experts and Diplomatic Stasis                                  1

   *Ms Natallia Khanijo*

2. Combating Cybercrime                                                     12

   *Major General Ashish Ranjan Prasad, VSM*

3. Challenges and Prospects of Cyber Security in the
   Indian Context                                                          21

   *Lieutenant General Nitin Kumar Kohli, AVSM, VSM*

4. Cyber Non-State Actors: The Cyber Taliban                               36

   *Colonel Sanjeev Relia*

5. Living With Cyber Surveillance and Espionage                            44

   *Colonel Sanjeev Relia*

6. Cyber Warfare and the Laws of War                                       53

   *Wing Commander UC Jha, PhD (Retd)*

7. Understanding Cyber Weapons                                             62

   *Colonel Sanjeev Relia*

8. Cyber Warfare & National Security Strategy                              70

   *Shri PV Kumar*

9. Cyber Weapons – The New Weapons of Mass
   Destruction?                                                            85

   *Lieutenant General Davinder Kumar, PVSM, VSM and Bar*
   (Retd)

10    Cyber Space, Outer Space and Information Space
      as the Non-Linear Strategic Frontiers                    99

      *Lieutenant General Davinder Kumar, PVSM, VSM and Bar*
      (Retd)

# Preface

The United Service Institution of India (USI) has been publishing its quarterly Journal since 1872. The Journal publishes articles on National Security, Military History and Defence related issues. To provide readers an insight into India's evolving national interests, geopolitical developments in the strategic neighbourhood, conflict spectrum, measures for developing comprehensive national power and defence capability, the USI decided to bring out its Annual Strategic Year Book in 2016. The articles for the Journal and the Strategic Year Book are contributed by eminent research scholars who have expertise on the subject. With the fast changing Security environment, some of the issues which have undergone rapid changes are covered in subsequent issues, at times without any linkage to the previous issues on the subject.

It was therefore consciously though that if these articles on similar themes are brought together in an edited book form, it will provide an opportunity for the readers to link the previous articles as a background on the subject and how the things shaped with times. It will also provide the readers to get to know the mindset of different authors on the same subject and provide a different line of thinking.

Cyber Warfare is gaining special significance in today's national security aspects. This book covers articles on Cyber Security, Cyberspace, Cyber Surveillance, Cyber Warfare including Laws of War, Cyber Weapons as new weapons of mass destruction et. al.

I am sanguine that this book will provide a good platform for those who are pursuing this subject for further research work.

– Editor

# Norms in Cyberspace: United Nations Group of Government Experts and Diplomatic Stasis[*]

*Ms Natallia Khanijo*

## Introduction

There has been a seismic shift in technological advancement in the last few decades. The proliferation of cyber networks, and their resultant impact on Information and Communication technologies (ICT), has pervaded almost every aspect of human functioning. The recent shifts in ICT functioning and the emergence of increasingly interconnected cyber networks have reduced metaphorical distances across the globe and potentially changed the ways in which nation states conceptualise their role in an increasingly hyper capitalised, multicultural, global order. The current framework of legality and ethical norm enforcement – by its very essential emphasis on lived exchanges in real time – is ill-equipped to deal with the hyper reality of alternative spatial and temporal constructs of existence. There is a need, therefore, to construct alternative methodologies of applying available normative/regulatory frameworks onto cyber discourse. The construction of such a framework is linked to the need for "including increased predictability, trust and stability in the use of ICTs, hopefully steering states clear of possible conflict due to misunderstandings. Additionally, norms [can also be seen] as guiding principles for shaping domestic and foreign policy as well as a basis for forging international partnerships."[1]

To this effect, several global bodies have been constituted aimed at multilateral, multinational and multi-stakeholder based 'regulation' of cyberspace. These include the creation of transnational forums for diplomatic

* This article was first published in the *Journal of the United Service Institution of India*, Vol. CXLVII, No. 609, July-September 2017.

dialogue such as the United Nations Group of Governmental Experts (UN GGE), the International Telecommunications Union, the Internet Governance Forum, the Shanghai Corporation Organisation, the Tallinn Manual, etc. whose primary motive is the theorisation, collaboration and regulation of norms and laws concerning Cyberspace. Currently, at the international level, at least 19 global and regional organisations are actively involved in the security and governance of the cyberspace. One of these bodies is the UN GGE instituted to deliberate on the 'Developments in the Field of Information and Telecommunications in the context of International security'. The UN GGE had its latest meeting over the course of 2016-2017 but due to the inability to conclude with a consensus, the expert body has been unable to release a consolidated report regarding the application of International Law to cyberspace. The lack of concrete norm formation and regulatory security architecture for an interconnected cyberspace is difficult to envision due to the amorphous nature of the realm itself. The ease of access to cyber technology, and the versatile nature of emergent threats – 'Lone Wolf' terrorists, 'Black Hat' hackers, non-state actors, geopolitical rivalries – cumulatively remain at the edge of transgressing State thresholds and the creation of the GGE was aimed at navigating this terrain of militarised cyberspace and infringement retaliation. This article attempts to examine the functioning of norms in cyberspace, the UN GGE as a process and specifically India's functioning with respect to the GGE, the reasons for its failure and what might potentially lie ahead.

## Norm Cycles and the UN GGE

The creation of a Norm Cycle for Cyber Discourse is primarily overseen by the United Nations. The debate concerning the emergence of ICTs and their impact on State sovereignty had first been introduced in the UN General Assembly (UNGA) regarding the field of Information and Telecommunication. As Roxanne Radu states, "In what concerns security in the cyberspace, three resolutions have been on the agenda. The First Committee of the UNGA discussed the resolution on 'Developments in the field of information and telecommunications in the context of international security' on a yearly basis starting in 1998; the Second Committee of UNGA discussed the 'Creation of a global culture of cybersecurity and the protection of critical information infrastructures', introduced in 2002 and adopted in 2005, and 'Creation of a global culture of cybersecurity and taking stock of national efforts to protect critical informational infrastructures', adopted in 2010".[2]

Two key bodies that have been linked to the creation of a Norm Framework have been the UN established Group of Governmental Experts which has served as the theorising body debating the modalities involved in the establishment of a Norm framework; and the International Telecommunications Union that is primarily concerned with the implementation of Norm Frameworks. The GGE emerged as a result of Russia's first proposition in 1998, regarding the establishment of a Group of Governmental Experts, who could examine the issue of Information Security. The General Assembly passed a resolution in 2002 concerning the "Creation of a Global Culture of Security",[3] and it outlined nine important elements that needed to be followed before engaging in the process of norm emergence. These elements are "awareness, responsibility, response, ethics, democracy, risk assessment, security design and implementation, security management and reassessment."[4] Furthermore, the Assembly also outlined the need for a resolution that determined the elements aimed at protecting 'Critical Infrastructure'. In its 11-point recommendation list, (that included Confidence Building Measures (CBMs), emergency/crisis communication networks, training exercises etc.), the assembly's resolution extrapolated on a lot of concerns that would form the basis of the various forums examining Cyber Discourse over the years.

## Iterations of the UN GGE

There have been five iterations of the UN GGE thus far. These have taken place in 2004, 2009, 2011, 2013 and 2016-2017 respectively. India has been involved with every single one except for the 2013 version. The first GGE was set up in 2004 by the First Committee[5] of the United Nations, however given the drastically divergent perspectives,[6] a consensus regarding the need for a normative framework could not be reached. The report concluded by saying that "given the complexity of the issues involved, no consensus was reached on the preparation of a final report."[7]

The second GGE was held in 2009 and there was a shift in global perception following the distributed denial of service (DDoS) attack on Estonia in 2007. While the constituent members were the same, there was a drastic change in the power discourse. The US stance regarding Cyber Discourse had altered in the interim and there was finally a consensus of sorts surrounding the need for a Cyber Security Architecture. The 2010 report concerning the proceedings, codified and extrapolated on a lot of the basic elements aimed at securing cyberspace. These measures included the need to identify malicious actors/victims/vulnerabilities and threats. The report concluded with five recommendations, namely :-[8]

(a) State Dialogue

(b) Implementing CBMs

(c) Information Exchange

(d) Capacity Building

(e) Clarification of Terminology describing Cyberspace

The third GGE took place between 2012-2013, with the intent to carry forward the discussion that began with the 2009 GGE and its 2010 report. The mandate for this third iteration was the need to examine potential threats in the realm of information security and collaborate on cooperative mechanisms to address the dangers of 'transnational anarchy'. The Group submitted its report in June 2013, wherein it made several recommendations and continued with the five-point agenda. Two major points that emerged were:-

(a) Application of International Law to Cyberspace. A breakthrough recommendation, this was one of the first concrete steps towards the establishment of a security architecture dealing with Cyberspace.

(b) Maintenance of human rights and fundamental freedoms.[9]

The fourth iteration of the GGE that took place in 2013 increased the membership from 15-20 states. The recommendations laid out by the Group followed the pattern of the previous two GGEs of creating a peaceful ICT environment through the establishment of secure sustainable architectures, protecting ICTs from National Security Adviser (NSA) intervention, implementing CBMs, etc.[10] Further the group also observed:-

(a) A need to follow international law while also recognising the State's right to take measures to secure its critical infrastructure keeping in mind four legal principles, i.e. humanity, necessity, proportionality and distinction.

(b) The question of attribution of blame and sustainability of evidence was also raised given the amorphous nature of cyber crime. The group noted that mere geographical indicators of State infrastructure being used to perpetrate malicious activities might be insufficient evidence as the State might be a victim as well. The need for substantiation to avoid wrongful condemnation on circumstantial evidence was also raised.

The GGE was working towards the establishment of a normative framework that could capitalise on mechanisms that were already in place in order to regulate inter and intra-state cyber behaviour to prevent the escalation of conflicts. As Roigas and Osula state "The text clarifies that the UN GGE is seeking 'voluntary, non-binding norms for responsible State behavior' that 'can reduce risks to international peace, security and stability."[11] The question that emerges therefore, is what went wrong? And, why did the GGE devolve into its erstwhile state of fragmented partisan politics?

## The 2016 GGE

The 2016 GGE was purported to deal with the impasse of norm applicability, multi-stakeholderism, legality and the militarisation of cyberspace. It was "tasked by the UN General Assembly with the study of existing and potential threats in the sphere of information security and measures to address them, including norms, rules, and principles of responsible behaviour of states, confidence-building measures, and capacity-building."[12]

While the earlier reports merely took note of the GGE proceedings, the 2015 report called upon Member States to follow the recommendations in their use of ICTs. Furthermore, while debating future topics of research and reference, the group also stated that "The United Nations Institute for Disarmament Research, which serves all Member States, is one such entity that could be requested to undertake relevant studies, as could other relevant think tanks and research organisations."[13] What needs to be noted is the fundamental ideological disjunct between the United States and its allies on the one hand and Russia and China on the other regarding the creation of a cyber normative discourse. The former were primarily keen on setting up principles to form a structure that would streamline the implementation of International Law of Cyberspace – including but not limited to the Laws of Armed Conflict and International Humanitarian Law which would inevitably lead to a kind of militarisation of cyberspace. Russia and China on the other hand, were more interested in protecting State sovereignty and autonomy. The nail in the coffin for the expert body, interestingly, came from the Cuban representative who stated that "it would legitimise a scenario of war and military action in the context of ICTs."[14] The US representative proceeded to claim that this wasn't true and that such a stasis would undo the groundwork of consensus formation that had been formed thus far, but the lack of a consensus and consequently a resolution meant that the body was unable to come up with a concrete report regarding the navigation of cybernetic terrain and fell back onto the earlier impasse regarding problems of attributability, minimum credible force, and military retaliation. The key disagreement revolved around the question of self defence in cyberspace and the applicability of legal frameworks regarding the same. While the previous iterations had approved of the applicability of International Law of Cyberspace, "the right to self-defence as enshrined in Article 51 had been a source of heated debates in all of the sessions leading to their adoption."[15]

## India and the GGE

India has been a member of all the GGEs barring the 2013 one. India has played an important role in facilitating cooperation and bridging the divide between polar ideological stances – particularly so in the 2011 GGE. Furthermore, "India has also acknowledged the seminal 2015 GGE report, with its cyber norms being endorsed in the India-US Cyber 'Fact Sheet' that was released during Prime Minister Narendra Modi's visit to Washington DC".[16] The question of access and inclusion are constantly raised with regard to the GGE given the discourse of power that emerges out of the tension between information rich and information poor nations. The politics and intersections of inclusivity in norm formation processes need to be examined in more detail given the fact that ICTs in particular are not just individual tools of state functioning but indispensable global architectures with interstitial, multi-pronged consequences. Being a part of the 2016 GGE was seen as an opportunity for India to navigate the politically complex terrain between developed and developing countries, and demonstrate its commitment to the creation of a peaceful, non-intrusive, collaborative ICT architecture. Even though several theorists believe that this would be the last GGE.

## GGE Limitations

The 2016-17 GGE might possibly be the last meeting of the group, and it was primarily constituted on Russia's insistence. Over the years, the GGE has certainly made certain important changes in the discourse surrounding cyberspace and ICT usage, but it needs to be noted that "cyberspace is a singularly complex setting within which to understand and try to shape norms. The problem is not simply the nature of cyberspace, (although, acknowledging the unique characteristics of cyberspace is crucial when exploring norms in this realm). Rather, the challenge lies in the often overlooked nature of norms themselves and how their defining features render them especially difficult to decipher – and, by extension, to attempt to design – in the context of cyberspace."[17] While the lack of consensus regarding cybernorm formation is disappointing, it cannot be considered a surprise given the variant constructions of sovereign ICT frameworks that differ from State to State. The Cuban representative raised a valid point with respect to negotiating/implementing a norm framework in cyberspace when there was such a drastic imbalance of power among the constituent countries. While cyberspace cannot be conflated with geopolitical complexities, it cannot be divested from them either. It is precariously balanced on the cusp of traditional warfare and even manifests in espionage, low grade phishing attacks and other such information warfare tactics.

There are several key issues that emerge in the aftermath of the proceedings that are worth examining. Firstly, one major limitation of the GGE is the lack of inclusivity in its constituent body. While increased inclusivity is considered a problem given the statistical certainty that the larger the base of the group, the harder it might be to broker a unanimous agreement on practicable issues. The exclusivity paradigm of geographical rotation is not really an acceptable alternative either. The current discourse regarding cyberspace, norm formation and ICT security architectures, stem from a predominantly western discourse which is tremendously problematic given that these legislative frameworks affect everyone in a globally interconnected world. Furthermore, the problem of inclusivity is twofold. Not only is there a problem with the horizontal axis of cyber discourse – wherein the academic predominance of the West stems from an inherent advantage in terms of access and technological superiority; but there is also a problem with the vertical axis of cyber norm formation wherein any constitutive body needs better representation at the level of the individual, private stakeholder and the country.

A legislative framework that might potentially constitute global norms with far reaching effects needs adequate representation from all stakeholders involved for the sake of ensuring that every single concern is engaged with.

The need for a more inclusive set of discussants as well as the need for multi-stakeholderism in an increasingly globalised world order is something that needs to be considered. There are several other bodies, treaties and groups attempting to pursue research in cyberspace and affect a secure architecture. Some key bodies are the International Telecommunications Union, the Internet Governance Forum, the International Committee of the Red Cross, the Shanghai Corporation Organisation, the United Nations Institute for Disarmament Research, etc.

It would be foolish to assume that geopolitical frameworks would not colour a country's approach to the implementation of cyber frameworks. As the Cuban representatives point out "an endorsement of the 'right to self-defense' [would] undermine asymmetric advantages which States that do not enjoy conventional superiority over their adversaries may have in cyberspace. So, Russia, which may be concerned that the United States will retaliate conventionally in response to a cyber operation that it deems to be an armed attack, would have concerns about including the phrase. On the other hand, India, which would want the option to respond to Pakistan's cyber operations through conventional means, may welcome the express affirmation of a right to self-defense."[18] This bias is intrinsically tied to complicated issues of deterrence in cyberspace and the establishment of retaliatory thresholds that vary from one geopolitical situation to the other. The variability of contexts, the subjectivity of thresholds and the anonymous/amorphous nature of the threat all collusively point towards a volatile and unstable geopolitical scenario which could become a hotbed for escalatory conflict on the basis of a country's interpretive retaliatory action. These scenarios do not even take into account the question of rogue states and non-state actors all of whom would lie outside the purview of global norm formation but possess the power to destabilise any fragile consensus that might be established.

The major issue of attrition and culpability remain unresolved as there is no established definitive understanding of the key terms of cyber norm formation. While there are theories of cyber deterrence, variant definitions of threats/actors, there is no consensus regarding mechanisms of attrition or investigative mechanisms that can be employed in these scenarios. Furthermore, as mentioned earlier, given the ease of access to cyber technology, and the relative ease with which attacks can be carried out and blame misdirected, there needs to be a concrete system in place that can deal with such dangerous liabilities without infringing on personal rights.

Keeping all these factors in mind, it's not surprising that the UN GGE reached such an impasse. The various other international bodies that exist need to collaborate towards addressing the key insecurities that permeate the amorphous fabric of cyberspace and contextualise threats in a systematic manner that is inclusive, equitable, consensus driven and maintains global peace.

## Conclusion

Totalitarian frameworks would be ill-equipped towards dealing with cybernetic transgressions and current legal architecture cannot just be placed onto cyberspace without modification and engagement. There is a need to reconfigure our epistemological frameworks to create a new sociological and geopolitical theory of knowledge regarding cyberspace and then work towards the implementation of particularised norms, tailored towards the specific contours of cyberthreats and cybernorms. There are several institutions and research organisations that attempt to do so such as the Tallinn Manual, that "address two subjects – the *jus ad bellum,* which regulates the use of force by States, and the *jus in bello,* the law that governs how States may conduct their military operations during an armed conflict and provide protection for various specified persons, objects, and activities."[19] While the Manual is not a legal document that is enforceable, it nevertheless provides an overview of potential ways in which Legal frameworks can be collated with cyber architecture. Compiled by lawyers and academics, the Manual provides a welcome first step towards engagement with the issue, but the levels and layers of inclusivity remain limited. True engagement with the complications of cyberspace would require re-engagement with the geopolitics that drives it as well. One cannot theorise the construction of cybernorm formation without examining the geopolitical realities within which it exists. Furthermore, given the rapid pace of technological proliferation, and the increasing vulnerabilities that are being capitalised on by rogue actors – such as the Wannacry ransomware attack and the Petya attack - it is absolutely essential that earnest measures towards cyber collaboration begin as soon as possible to prevent the devolution of the geostrategic world order into an apocalyptic cyber wasteland.

# Endnotes

1   Osula, Anna-Maria, O. Rõigas, *International Cyber Norms Legal, Policy & Industry Perspectives.* CCDCOE, NATO CCDCOE Publications, Tallinn 2016.

2   Radu, Roxanne, Power Technology and Powerful Technologies: Global Governmentality and Security in the Cyberspace. In J.-F. Kremer, & B. Muller, *Cyberspace and International Relations,* Berlin: Springer, 2016.

3   Resolution - 57/239 Creation of a global culture of security, adopted by the General Assembly, 2002, Available at https://www.oecd.org/sti/ieconomy/UN-security-resolution.pdf, Accessed on 06 Sep 2017

4   Radu, Roxanne, Power Technology and Powerful Technologies: Global Governmentality and Security in the Cyberspace. In JF Kremer, & B Muller, Cyberspace and International Relations, Berlin: Springer, 2016.

5   Disarmament and International Security.

6   Russia and China disagreed with the United States' ideas as they believed those ideas would lead to the militarisation of Cyberspace.

7   Report of the Secretary General, United Nations, Submitted by the Group of Governmental Experts on Developments in the Field of Information and Telecommunications in the Context of International Security, 05 Aug 2005, Available online at https://disarmament-library.un.org/, Accessed on 09 Sep 2017.

8   Report of the Secretary General, United Nations, 30 Jul 2010, Available online at http://www.unidir.org/files/medias/pdfs/information-security-2010-doc-2-a-65-201-eng-0-582.pdf

9   As set forth in the Universal Declaration of Human Rights and other international instruments.

10  Ibid.

11  Osula, Anna-Maria, O Rõigas, *International Cyber Norms Legal, Policy & Industry Perspectives.* CCDCOE, NATO CCDCOE Publications, Tallinn 2016.

12  Korzak, E (2017). *UN GGE on Cybersecurity: End of an Era.* Available at The Diplomat: http://thediplomat.com/2017/07/un-gge-on-cybersecurity-have-china-and-russia-just-made-cyberspace-less-safe/ Accessed on 10 Sep 2017.

13  Report of the Secretary General, United Nations, 22 Jul 2015. Available online at http://www.un.org/ga/search/view_doc.asp? symbol=A/70/174&referer=http://giplatform.org/actors/united-nations-group-governmental-experts-developments-field-information-and&Lang=E Accessed on 08 Sep 2017.

14  Korzak, E (2017). UN GGE on Cybersecurity: End of an Era. *Available at The*

*Diplomat:* *http://thediplomat.com/2017/07/un-gge-on-cybersecurity-have-china-and-russia-just-made-cyberspace-less-safe/ Accessed on 10 Sep 2017.*

15  Michael N Schmitt, Vihul Liis, (Ed.), *Tallinn Manual 2.0 on the International Law applicable to Cyber operations,* Cambridge University Press, (2017)

16  Sukumar, UN Reconstitutes its Top Cyber Body, This Time with India at the High Table, 2016, Available at The Wire: https://thewire.in/44696/un-reconstitutes-its-top-cyber-body-this-time-with-india-at-the-high-table/ Accessed on 09 Sep 2017.

17  Osula, Anna-Maria, O Rõigas, International Cyber Norms Legal, Policy & Industry Perspectives. *CCDCOE, NATO CCDCOE Publications, Tallinn 2016.*

18  Sukumar, A M, *The UN GGE Failed. Is International Law in Cyberspace Doomed As Well?* 2017, Available at Lawfare: https://lawfareblog.com/un-gge-failed-international-law-cyberspace-doomed-well Accessed on 08 Sep 2017.

19  Michael N Schmitt, Vihul Liis, (Ed.), *Tallinn Manual 2.0 on the International Law applicable to Cyber operations,* Cambridge University Press, (2017).

†  **Ms Natallia Khanijo** has done her graduation from Lady Sri Ram College and post-graduation from Miranda House, University of Delhi. Currently, she is researching on 'Cyber Issues' in Institute of Defence Studies and Analyses (IDSA), Delhi.

# Combating Cybercrime[*]

## Major General Ashish Ranjan Prasad, VSM

### Introduction

The growth of internet has been the biggest social and technological change of our lifetime. It is a great medium that allows people across the world to communicate and has become increasingly central to our economy and society. But the growing role of Cyberspace has also opened up new threats from Cyber criminals along with new opportunities. The high degree of anonymity, speed of communication, efficiency and reach to the masses has led to it being exploited by Cybercriminals. Therefore we should have a clear cut approach towards handling of Cybercrimes at national level both at organisational and individual levels. The Government should be in a position to ensure protection of the organisations and individuals from crime, fraud and identity theft etc.

### Categories of Cybercrimes

Criminals from all corners of the globe are already exploiting the Internet to target individuals and organisations. Few main categories of Cybercrimes can be described as below :- [1]

(a) **Breaking into Communication Services.** Unauthorised access of information services compromises security.

(b) **Promoting Criminal Activities.** Cyber domain is being used extensively to facilitate organised drug trafficking, gambling, money laundering and arms smuggling. The use of encryption

* This article was first published in the *Journal of the United Service Institution of India*, Vol. CXLVI, No. 603, January-March 2016.

technology places criminal communications beyond the reach of law enforcement.

(c) **Cyber Piracy.** The temptation to reproduce copyrighted material for personal use, sale or free distribution violate anti-piracy laws and are treated as criminal offences.

(d) **Cyber-Stalking.** Computer systems can also be used for harassing, threatening or intrusive communications, by means of "cyber-stalking".

(e) **Financial Irregularities and Tax Frauds.** Hi-tech online transactions over secured channels cannot be tracked with traditional countermeasures.

(f) **Electronic Vindictiveness and Extortion.** Dependence on complex data processing and telecommunications systems is prone to damage or interference by electronic intruders.

(g) **Investment and Marketing Frauds.** The increasing use of internet marketing and investment allow fraudsters to enjoy direct access to millions of prospective victims around the world, instantaneously.

(h) **Electronic Eavesdropping.** Remote monitoring of computer radiation and eavesdropping compromises information security.

(i) **Electronic Funds Transfer Fraud.** Digital information stored in credit card can be counterfeited and misused.

(j) **Identity Theft.** Identity theft is used by Cybercriminals for monetary gains and serves as a gateway to other Cybercrimes such as tax-refund fraud, credit-card fraud, loan fraud and other similar crimes.

(k) **Theft of Sensitive Data.** Sensitive information related to government, organisations or individuals attract the attention of Cybercriminals.

## Cybercrime – Impact on National Security

Use of Cyberspace in civil as well as military domains has today become an intricate component of national power. With defence forces adopting more complex Information and Communication systems and upgrading to network centric warfare, they are at higher risks of cyber-attacks. The

13

"Make in India", "Digital India" and "Smart Cities" are flagship programmes with a vision to transform India into a digitally empowered society, foster innovation, knowledge economy and infrastructure development in India by leveraging the use of information technology. It goes without saying that this accelerated capacity building has enormous implications for the Country's cyber-security posture. At the same time, threats from both state and non-state actors are weakening the very foundations of these concepts. None of the existing international laws on cyberspace apply to the terrorist organisations who have adapted themselves in innovative ways to become one of the most ardent users of cyberspace for a variety of criminal activities - from communication, to finance, as well as for recruitment, networking and psychological operations (Psy Ops) as we are currently witnessing. As the visual and real worlds get increasingly integrated with the Internet of Things (IoT), it is only inevitable that use of cyberspace for destructive purposes will pose a serious threat to national security.

## Challenges in Handling Cybercrimes

The human society around the world is racing ahead with innovative trends in information technologies. This has also given rise to well managed criminal activities where the commodity, personal information or data moves far too quickly for conventional law enforcement methods to keep pace. Detecting, quantifying and preventing Cybercrime is a difficult task. A few challenges are as under:-

(a) The Cyberspace is not limited by well-defined boundaries and hence the actions in the Cyber domain cannot be traced to the source of origin. These features are being exploited by non-state actors for perpetration of misdemeanors in the Cyber domain.[2]

(b) The reach and complexity of the offences committed in the Cyber domain are continually on the rise thereby affecting the Government as well as the institutions and individuals.

(c) As the volume and value of information hoisted in the electronic domain have increased, innovative methods are being adopted by Cyber criminals as more convenient and profitable ways of carrying out their activities anonymously are being evolved.

(d) The ability of adversaries to produce, distribute and utilise malicious code with ease maximises their gains and at the same time pose challenge to threat evaluation and traceability.

(e) Targeted attacks are growing faster, stealthier, multifaceted and extremely difficult to analyse and are causing risk to national security.

## Current Scenario at National Level

With the increase in frequency of Cybercrimes in India and registration of Cybercrimes showing an annual quantum jump over the past years, an expert group set up by the Home Ministry has suggested setting up of a dedicated body which is proposed to be called Indian Cybercrime Coordination Centre (I4C). This will facilitate online reporting of Cyber offences, apart from monitoring, analysing and countering these new-age crimes. This national body will have linkages with state police and will e-integrate around 15,000 police stations across the Country, and NatGrid. This dedicated body will have high-quality technical experts and R&D experts to develop cyber investigation tools to coordinate the aforesaid actions. Also, the body can take up long-term training programmes for the law enforcement agencies and even judiciary on investigation and prosecution of Cybercrimes.

The proposed I4C will have real-time analytics of Cybercrime along with their types. This will help strengthen India's case in seeking cooperation from global Internet firms having servers abroad, to tackle various types of Cybercrimes. Also the planned architecture should have routing of the Internet services through a single, common gateway rather than separate gateways now used by the Country's Internet service providers. There is also need to have a relook at the legal framework, including the Indian Evidence Act, 1872 and Information Technology Act, 2008 against any existing loopholes or voids to deal with Cybercrime.

In view of emerging challenges in the Cyber world and spiralling Internet crime rate, State governments also need to take stern measures. There is a need to create nodal centre for effective policing of social networking sites and anti-terror activities in Cyberspace. All types of Internet related activities ranging from virtual policing, automated threat intelligence, Cyber forensics and tracking system need to be put in place.

There is a need for security compliance and a legal system for effective dealing with internal and external Cyber security threats. India needs good coordination between law and technology to come out with a mechanism of cooperation among states, agencies and countries to address these challenges. The strategy and roll-out plans are needed for addressing the challenges related

15

to Cybercrime in the short-term and the mid-term, with a mechanism to review the same on a long-term basis. In addition to the existing mechanisms, a strategy needs to be promulgated which states the vision, objective and approach for Cybercrime prevention in India. For this purpose, the Indian Government has set up its own 'Cyber Security Architecture' comprising following bodies :-[3]

    (a)  National Cyber Coordination Centre (NCCC).

    (b)  National Critical Information Infrastructure Protection Centre (NCIIPC).

    (c)  Grid Security Expert System (GSES).

    (d)  National Counter Terrorism Centre (NCTC).

    (e)  Cyber Command for Armed Forces.

    (f)  Central Monitoring System (CMS).

    (g)  National Intelligence Grid (NATGRID).

    (h)  Network and Traffic Analysis System (NETRA).

    (i)  Crime and Criminal Tracking Network & Systems (CCTNS).

## Cybercrime Early Warning, Reporting and Response

Cybercrime, like any other crime, should be reported to appropriate law enforcement authorities depending on the scope of the crime. Quick access of such reporting system should be made available to victims. Law enforcement authorities should be made aware online about the suspected criminal or civil violations. Maintaining centralised database will provide a repository to law enforcement and regulatory agencies at the national, state and local levels. The activities needed to be pursued under this initiative include :-[4]

    (a)  Adopting and deploying state-of-art tools and techniques.

    (b)  Creating a structured knowledge repository.

    (c)  Strengthening partnership and cooperation with industry, international Computer Emergency Response Team (CERTs) and security forums.

(d) Acquisition of intelligence about vulnerabilities, threats, and security risks collated from a comprehensive list of sources.

(e) Establishing a collaboration platform for engaging with security community.

## Legal Architecture

Cybercrime raises several challenges for traditional criminal law and the criminal justice system in general.[5]

(a) The first challenge is to define the types of Cybercrimes and include the same in its conceptual framework for influencing national legislation on Cybercrime and policies at international level.

(b) The second challenge is that the Information and Communication Technology (ICT) is complex and dealing with crime involving these devices requires well-trained personnel in the investigation phase, during prosecution, and in courts.

(c) As a third challenge, many Cybercrimes occur in virtual environments like mobile phone channels or the Internet. This feature frequently clashes with the main operational criteria of the criminal justice systems, namely sovereignty and the territoriality principle, hence, it requires countries to establish clear rules on a legal system's jurisdiction over these offences.

(d) The fourth challenge is that the world of ICT moves at a pace different from that of physical world. Crimes occur in a fraction of a second and may spread with astonishing speed.

(e) Lastly, the challenge due to virtual nature of Cybercrimes wherein a perpetrator may be in a different jurisdiction from the victim and the legal definitions of the criminal behaviour in the two legal systems may not match.

Law enforcement agencies must, therefore, take rapid action for collecting and preserving the digital evidence for use in criminal proceedings. If criminal justice systems are to deal effectively with these problems relating to the repression of Cybercrime, they must update their legislation and law enforcement systems where these are unable to cope with investigation and prosecution of the phenomenon. Successful policies undertaken by the foreign countries may be adopted for better utility against Cybercrimes.

## Cybercrime Prevention (R&D, Training and Awareness)

There are huge gaps in the number of trained Cyber security professionals available in the Country as compared to the overall requirements. R&D in Cyber security is unsatisfactory. Non-availability of proficient Cyber experts within law enforcement agencies and inappropriate implementation of the strategy means that very few measures are in place to immobilise a larger set of Cyber sleuths to counter the menace of Cybercrime. Additionally, in order to identify the modus operandi of the criminals, it is essential to understand the psychology rather than just relying on tools and technology.

Spreading awareness on Cybercrime prevention is an essential requirement. The Cybercriminals are constantly seeking new ways to attack and identify potential victims. In recent times, critical infrastructure of a few countries was successfully penetrated due to the low awareness level of most users, through phishing and social engineering methods.

Citizen awareness programmes should be launched to prevent Cybercrimes, as proactive mitigation has to be achieved through multiple media channels. Mechanisms should be established for independent monitoring of awareness programmes at regular intervals to evaluate the number of people and regions covered through the awareness programmes. Awareness material should be updated regularly as well.

## International Collaboration

Since the Cyber world transcends all physical barriers, and is also being transnational in nature, it is but obvious that nations across the globe need to strengthen their cooperation and form alliances as well as ensure that their legal, technical and institutional measures are put in place. Though the IT Act, 2008 categorises Cyber offence as a crime in India; it has its own limitations; thus, it lacks the necessary execution on ground. This includes investigation, prosecution and consecutive extradition of a foreign national as well.[6]

India remains a non-signatory to the Budapest Convention, which is the international treaty seeking to address Cybercrime by harmonising national laws, improving investigative techniques and increasing cooperation among nations. It will be beneficial to have collaboration with International Cyber Security Protection Alliance.

## Summary of Action Plan

A summary of action plan which needs to be initiated at the national level is given as below :-[7]

(a) **National Response.** Improve our detection and analyses capabilities to defeat high-end threats, with a focus on the critical national infrastructure.

(b) **Governance.** Establish internationally agreed 'rules of the road' on the use of Cyberspace and ensure its implementation.

(c) **Security.** Manage and ensure that the key critical infrastructure remains safe and resilient.

(d) **Cooperation.** Share information of threats in Cyberspace, including from private sector, for creation of security database at national level.

(e) **Execution.** Enable all law enforcement agencies to handle Cybercrimes and forensics.

(f) **Reporting and Response.** Build an effective chain of reporting Cybercrime and improving the police response at local level for those who are victims of crime.

(g) **International Synergy.** India should ratify all international forums so that Cybercrimes can be prosecuted across borders and offenders are denied safe havens and offshore help.

(h) **Legal Framework.** Courts of Law should be empowered with enforcement capabilities to report, react, disrupt and prosecute Cybercrimes.

(i) **Core Competence.** Promote development of a cadre of skilled Cyber security professionals to retain an edge in the area of crucial key skills and technologies.

(j) **Awareness.** As prevention is a key, we need to work to raise awareness, educate and empower people and firms to protect themselves online.

(k) **Role Model.** Model the best practices on Cyber security in the Government's own systems thereby setting up strong standards for suppliers to the government agencies.

## Conclusion

To positively impact the Cyber security ecosystem and to combat Cybercrime, it is imperative that efforts and resources are dedicated to operationalise the Nation's Cyber security strategy. If such initiatives are driven from the highest level of the Government, it ensures that all stakeholders are interested and engaged in contributing to the success of initiatives or programmes. Such commitment alone, though it is an important enabler, is not sufficient to guarantee the success of an initiative or programme. Monitoring and review mechanisms are essential to analyse and assess progress as well as to consider measures for re-calibration and course correction as may be required. It is important to define milestones and operationalise the strategy as per the desired impact of the initiatives.

## Endnotes

1   The UK Cyber Security Strategy: Protecting and Promoting the UK in a Digital World, Nov 2011, CRET-UK.

2   ITU National Cybersecurity Strategy Guide Dr. Frederick Wamala (Ph.D.), CISSP, Geneva, Switzerland.

3   Strategic national measures to combat Cybercrime: Perspective and Learning's for India, August 2015, by ASSOCHAM India. p 10 – 12.

4   XII Five-Year Plan on Information Technology Sector, Report of Sub-Group on Cyber Security, Ministry of Communications & Information Technology, Department of Information Technology, Government of India. p 111.

5   Op. Cit 3.

6   National Cyber Security Policy -2013, Ministry of Communication and Information Technology, Department of Electronics and Information Technology, Govt of India.

7   Ibid.

---

† **Major General Ashish Ranjan Prasad, VSM** was commissioned into the Corps of Signals on 13 Jun 1981. He commanded 14 Corps Operating Signal Regiment and 2 Signal Group (Electronic Warfare). Presently, he is posted as the Additional Director General Signal Intelligence at the Integrated HQ of MoD (Army).

# Challenges and Prospects of Cyber Security in the Indian Context[*][†]

## Lieutenant General Nitin Kumar Kohli, AVSM, VSM

### Introduction

The World Telecommunications Day is celebrated on 17 May. India is marching towards Digital India. Approximately 278 million people out of a population of 1.2 billion are connected on the internet.[1] Out of six lakh villages, mobile connectivity has been provided to 5.5 lakh villages, only 50,000 villages are left. Government's endeavour is to connect all citizens of the Country through digital means. If this dream has to be realised, we need policies, techniques and procedures which address the issues of 'Cyber Risk and Security' to guarantee success to these concepts.

Indian Cyberspace is under constant threat. Just to highlight the gravity of the situation, according to data from the Computer Emergency Response Team – India (CERT-IN), the cyber espionage incidents have gone up from 23 reported incidents in 2004 to a mammoth figure of 62,189 in 2014.[2] Given the number of critical systems reachable via the Internet coupled with the growing technological advancement of other countries and our heavy reliance on imported hardware and software, 'It's a question of when, not if.'

### Shifting Trends

From worms and viruses to Distributed Denial of Service (DDoS) and Advanced Persistent Threats (APTs), in the past quarter of a century the

---

* Text of the talk delivered at USI on 20 May 2015 with Lieutenant General Davinder Kumar, PVSM, VSM (Retd) in Chair.

† This article was first published in the *Journal of the United Service Institution of India*, Vol. CXLV, No. 600, April-June 2015.

sophistication, impact and scale of cyber-attacks have evolved significantly. Technologically advanced nations have been developing ways to use information as a weapon and target financial markets, government computer systems and utilities. Some of the global prominent attacks are Stuxnet, Flame, Dark Seoul and Sony Pictures Entertainment Hack. These attacks were carried out using espionage and a combination of backdoors, Trojans and worms, and were state sponsored.

Although, the bigger attacks are reported, less noticed is growing cottage industry of ordinary people hiring hackers for much smaller acts of espionage. Websites like "Hacker's List" seeks to match hackers with people looking to gain access to competitor's e-mail accounts, databases etc.[3]

## Our Neighbours

**China.** It has been estimated that 90 per cent APTs are traced to China. China has been accused of cyber attacks not only on the US or India, but also across many nations of the world. China now has both the intent and capability to launch cyber attacks 'anywhere in the world at any time'. China has mounted almost daily attacks on Indian computer networks of both government and private sector, showing its intent and capability.[4] The Chinese are constantly scanning and mapping India's official networks. This is China's way of gaining 'an asymmetrical advantage' over a potential adversary.

**Pakistan.** China has found an ally in Pakistan whom it can use as a launch pad to inflict cyber attacks on India. On 26 Jan 2014, Pakistani hackers defaced 2118 Indian websites.[5] Pakistan may use mass media and internet to disturb the secularly balanced India by triggering religious sentiments of Indians. The Assam riots that triggered a widespread exodus of north eastern students from cities such as Bangalore and the widespread stone pelting incidents in J&K, confirmed the subversive games Pakistan plays through social networks.

## Indian Preparedness

India has issued Cyber Security Policy and legal framework to secure its Cyberspace as elaborated in the succeeding paragraphs.

## National Cyber Security Policy 2013

The objective of National Cyber Security Policy (NCSP) 2013[6] in broad terms is to create a secure cyberspace ecosystem and strengthen the regulatory framework. A National and sectoral 24x7 mechanism has been envisaged to deal with cyber threats through National Critical Information Infrastructure Protection Centre (NCIIPC). CERT-IN has been designated to act as a nodal agency for coordination of crisis management efforts. CERT-IN will also act as umbrella organisation for coordination actions and operationalisation of sectoral CERTs.

The policy calls for effective public and private partnership creating a think tank for cyber security evolution in future. Other important facets of the policy are promotion of research and development in cyber security, development of human resource through education and training programmes and creating a workforce of 500,000 professionals trained in cyber security in the next five years. The policy document aims at encouraging all organisations whether public or private to designate a person to serve as Chief Information Security Officer (CISO) who will be responsible for cyber security initiatives. The release of the NCSP 2013 is an important step towards securing the cyber space of our Country.

## Legal Framework: IT Act 2000 and IT Act (amended) 2008

The highlights are :-[7]

(a) **Definition of Computer System and Punishment for Cyber Offences.** Provides comprehensive definition of computer systems and ascertains liability on various types of crimes.

(b) **E-Governance and E-Transactions.** Provides legislation for E-governance & E-transactions.

(c) **Authority to Government.** Authorises government for interception, monitoring and blocking of websites.

(d) **Protected Systems.** Under the Act, critical systems can be declared as 'protected systems' and security breaches of such systems attract imprisonment.

(e) **Appellate Tribunals.** Cyber Appellate Tribunal, which is now operational, is expected to expedite legal proceeding of cyber crime cases.

## Challenges of Cyber Security in India

### Lack of Comprehensive Policy

The NCSP was issued in 2013 but has been proceeding in fits and starts. Some of the shortcomings are as given below:-

(a) **Need for a National Security Policy.** The National Security Council (NSC) has not published any official document outlining the National Security Policy (NSP). Since NCSP was not a subset of any NSP, it was relegated to the status of an isolated departmental document of the Ministry of Communication and Information Technology (MoC&IT) rather than desirable national level policy. The policy does not give any road map, timelines and funding for its implementation.

(b) **Insufficient Private Sector Input,** Including Public-Private Partnerships (PPPs). During the formulation of NCSP minimal effort was made to obtain input and expertise from other sectors. Although it engaged with industry groups such as the Federation of Indian Chambers of Commerce and Industry, the process was half-hearted at best. This excludes an entire pool of talent that is available from India's many start-up firms, as well as individuals.

(c) **Exclusion of Armed Forces.** Unlike the policies of cyber mature nations that recognise cyber security to lie at the broad intersection of both military and commercial networks, the NCSP is largely ambiguous about the role, interplay and interdependence of these two distinct aspects of national cyber security.

(d **International Cooperation and Advocacy.** The policy fails to mention the leadership role India should be playing in a variety of areas in cyber security, including development of international security standards, testing of ICT products, cyber security norms and conventions, solutions to the issues of Internet governance, among many others.

### Organisational Shortcomings

There are around six apex bodies, five ministries and almost thirty agencies that make up the cyber organisation.[8]

It requires serious introspection to make the entire structure conducive to effective command and control. It is recommended that GoI reconfigures apex bodies to create a single empowered authority to resolve the predicament of multiplicity at the top level.

## Lack of Internationally Accepted Policies and Laws

The biggest hurdle before curbing cyber threats at the international level is lack of harmonisation at international level. Till now we have no 'Internationally Acceptable Definition' of cyber warfare. Further, we have no universally acceptable cyber crimes treaty as well.

## IT Act 2000 and 2008

The provisions of IT Act are mostly bailable and there has been very low rate of convictions.[9] It will not be wrong to say that it is effective in metropolitan cities like Mumbai, Delhi, Hyderabad, Bhopal, Bangalore, etc, but it is feeble in tier-two cities as awareness of the law by enforcement agencies remains a big challenge. This needs to be suitably addressed.

## Supply Chain Integrity

Supply chain integrity has become paramount with the needle of suspicion pointing towards the hardware and software that make up the brains and body of cyberspace. While much of the equipment used in global networks is supplied by China, the storage and data storage networks are largely of the US companies. The dominance of Chinese companies like Huawei and ZTE with reportedly close links with the Chinese military is a matter of concern.[10] In addition to the widely reported issues with hidden backdoors and kill switches, it is also a fact that network equipment providers get access to sensitive information in the course of providing after sales support.[11]

## International Cooperation

The MoUs signed by India have been lopsided in favour of other nations. The Indo-US Strategic Dialogue held in June 2013 renewed focus on cyber security with the establishment of a Strategic Cyber Policy Dialogue of cyber experts. While the macro issues important to the US are being addressed through these dialogues, they do not seem to provide scope for addressing issues important to India such as evolving the necessary mechanisms for rapid information sharing in the law enforcement process.[12]

## Non Adherence to International Best Practices: ISO 27001

ISO 27001 certification is suitable for any organisation, large or small and in any sector for protection of critical information, such as in the banking, financial, health, public and IT sectors. All critical sector organisations under Central Government ministries/departments are mandated to implement information security best practices as per ISO 27001. However, there are only 546 organisations in the country which have obtained the certification. What is more intriguing is that the Department of Electronics and Information Technology (DeitY) has not made any effort to ascertain as to why all the Government organisations have failed to obtain ISO 27001 certification.[13]

**Large User Base with Few Experts.** With a population of around 1.21 billion, India has so far only 65,000 trained personnel pertaining to cyber security as against the estimated requirement of 5 lakh trained personnel. In addition, there are only 97 Master trainers and 44 empanelled auditors by CERT-IN in the country.[14]

**Pirated Software.** According to the Global Software Piracy Study done by an independent firm, Business Software Alliance (BSA), about 60 per cent of Indians used pirated software. Only 33 per cent of companies in India have written policies in place requiring use of properly licensed software. This increases the chance of encountering malware.

**Data Traffic Transit through Foreign Countries.** Much of the data traffic that traverses through cyberspace touches the US networks at some point, or is carried over these networks. Also, majority of the websites of commercial, NGOs, individuals and private organisations are hosted outside India and thus the data is always vulnerable.[15]

**Lack of Strong Security Culture.** India lacks a strong security culture. A country's security culture should permeate all those who are actively engaged in security-related sectors. This is especially important in the cyber security domain, where every individual has the potential to be both a defender and a victim.

## Cyber Balance Sheet of Cyber Mature Nations

### The USA

Till late nineties, the US suffered from various shortcomings like inadequacy of national policy, multiple organisations, wasteful funding and ineffective regulations to penalise the perpetrators.

**Policy Framework.** The National Security Strategy (NSS) released in May 2010 called for integration of various agencies.[16] As per the guidelines in NSS the Department of Defence (DoD) coined its cyber concerns in the National Defence Strategy (NDS) and the Quadrennial Defense Review (QDR).[17] Further, these strategic documents were used by the Joint Staff to formulate the National Military Strategy (NMS). Now, DoD has cyber policies at strategic, operational and tactical levels.

**Integration of Organisations.** The responsibility of cyber security was spread across the Department of Homeland Security (DHS), DoD and Department of Justice (DoJ), which worked in independent silos and failed to prevent cyber attacks against the US. In 2008 National Cyber Investigative Joint Task Force (NCIJTF) was formed and drove the US towards unity of command.

### US CYBERCOM

In the year 2006, Pentagon reported an all time high 360 million attempts, including hacking into the US $300 billion Joint Strike Fighter project.[18] The Pentagon spent nearly 14 months in 2008 cleaning the worm 'agent.btz' which originated from a DoD facility in the Middle East. Under these circumstances, the US formed the United States Cyber Command (USCYBERCOM) on 23 June 2009 under the US Strategic Command (USSTRATCOM).[19] A four star general wears a dual hat of Director, National Security Agency and Commander, USCYBERCOM. The Command is charged with putting together existing cyberspace resources, creating synergy and synchronising war-fighting effects to defend the information security environment.

### Other Cyber Programmes of the USA

Various programmes are run by the NSA with near impunity due to provisions and authorisations under Foreign Intelligence Surveillance Act (FISA). Some of the NSA's programmes are directly aided by national and foreign intelligence agencies as well as by large private telecommunications and internet corporations such as Verizon, Telstra, Google, Microsoft and Facebook.

The cyber security firm Kaspersky Laboratory has disclosed in Feb 2015 that a US cyber espionage group called the 'Equation Group' embedded surveillance tools on the hard drives produced by a number of well known manufacturers like Western Digital, Seagate, Hitachi and Toshiba. It was almost impossible to get rid of the malware, even after disk reformatting and re-installing the computer system.

The US DoD has declared its Cyber Strategy in Apr 2015.[20] This new strategy sets prioritised strategic goals and objectives for DoD's cyber activities and missions to achieve over the next five years. It focusses on building capabilities for effective cybersecurity and cyber operations to defend DoD networks, systems, and information; defend the nation against cyberattacks of significant consequence; and support operational and contingency plans. The strategic goals listed are as follows:-

(a) Build and maintain ready forces and capabilities to conduct cyberspace operations.

(b) Defend the DoD information network, secure DoD data, and mitigate risks to DoD missions.

(c) Be prepared to defend the US homeland and US vital interests from disruptive or destructive cyberattacks of significant consequence.

(d) Build and maintain viable cyber options and plan to use those options to control conflict escalation and to shape the conflict environment at all stages.

(e) Build and maintain robust international alliances and partnerships to deter shared threats and increase international security and stability.

## China

The Chinese People's Liberation Army (PLA) is actively developing a capability for computer network operations (CNO) and is creating the strategic guidance, tools and trained personnel necessary to employ it in support of traditional war fighting disciplines. The Chinese have adopted a formal IW strategy called 'Integrated Network Electronic Warfare' (INEW) that consolidates the offensive mission for both computer network attack (CNA) and Electronic Warfare.

## Organisations and Capabilities of PLA

Chinese efforts to dominate the information space are driven primarily by three goals – exercise control over their populace, dominate adversaries by dominating the information space and finally overcome the technological gap with the West through strategic intelligence acquisition in the cyber domain. The assessed Cyber structure of China is illustrated in the succeeding paragraphs.[21]

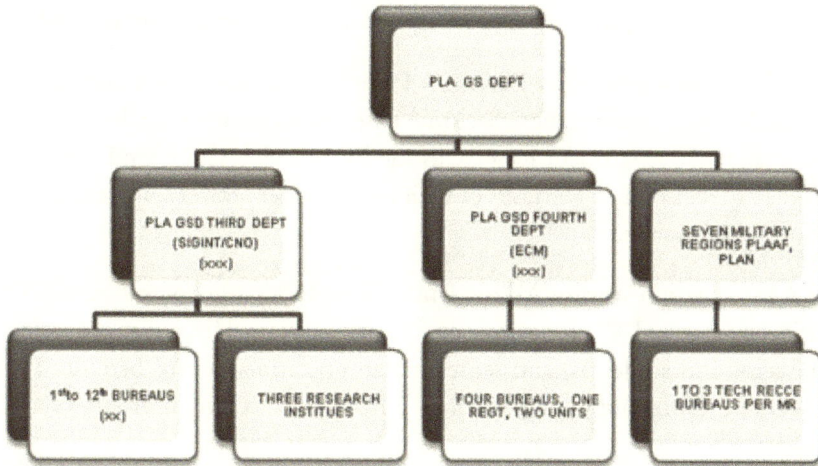

Cyber Organisation of PLA

## Cyber Organisation of PLA

**Third Department.** It is tasked with the foreign signals collection, exploitation and analysis as also communications security for the PLA's voice and data networks. The GSD Third department directly oversees following entities : -

(a) **Operational Bureaus.** There are twelve operational bureaus and every operational bureau has a specific mission such as radio or satellite communications interception, cryptology, translation, information assurance, intelligence analysis, cyber operations which include exploitation, defence and attack. The second, seventh and eighth bureaus are likely to be involved in Cyber operations.

(b) **Technical Reconnaissance Bureaus.** The PLA maintains at least six technical reconnaissance bureaus (TRB) that are responsible for SIGINT collection against tactical and strategic targets and have apparent CNO duties focussed on defence or exploitation of foreign networks.

(c) **Research Institutes.** Science and Technology Intelligence Bureau and, Science and Technology Equipment Bureau oversee three Research Institutes namely 56th Research Institute, 57th Research Institute and 58th Research Institute which focus on codes and passwords, development of communication intercepts and signal processing systems, cryptology and information security technology.

**PLA Information Security Base.** On 19 July 2010, the PLA is said to have established the "Information Security Base" headquartered under the PLA General Staff Department to serve as the PLA's Cyber Command. The base is likely to consolidate key tasks of China's computer network operations and information warfare.

**PLA Information Warfare Militia Units.** From about 2002 onwards, the PLA has been creating IW militia units comprising personnel from the commercial IT sector and academia, and represents an operational nexus between PLA CNO operations and Chinese civilian information security professionals.

**Hacker-State Collaboration.** The Chinese government demonstrates willingness to leverage the power of hacker communities by direct collaboration between state and hackers so that the CNO is coordinated and mission oriented with the ability to deny state involvement. Xfocus is one of the many such hacker groups which has transformed into a commercial information security company. "Red Hackers" or "Hongkes" are Chinese citizens, often motivated by patriotism or financial gains, who act as modern-day privateers attacking foreign targets.

**Cyber Security Measures.** Chinese authorities have instituted 'The Great Firewall' to regulate the internet in mainland China. Also, China has shifted to indigenous operating system based on Unix called Kylin. It also employs indigenously developed search engines and social networking websites. Some of these are given below: -

|     | Popular websites | Chinese Equivalent | Type of Application |
| --- | --- | --- | --- |
| (a) | Twitter | SinaWeibo | Mass Messaging |
| (b) | Facebook | Renren, Pengyou | Social Networking |
| (c) | Google Talk | QQ | Instant Messaging |
| (d) | MySpace | Douban, Diandian | Forum/Blog |
| (e) | Youtube | Youku | Video Sharing |
| (f) | Whatsapp | Wechat | Mobile Voice and Text App |
| (g) | Foursquare | Jiepang | Location-based Social Networking App |

## Recommendations

**Formulation of National Security Policy.** India should formulate an all-encompassing National Security Policy (NSP) and the National Cyber Security Policy should be a subset of this policy. Thereafter, National Cyber Doctrine and Cyber Security Strategy can be formulated by respective ministries. This would introduce tier-based 'policy-doctrine-strategy' formulation and ensure 'whole-of-nation' approach in cyber security. The policy should give the road map, timelines and funding for its implementation.

**Reconfigure Apex Organisation.** The apex bodies should be reconfigured to create a single empowered authority to resolve the predicament of multiplicity at the top level. It is proposed that an exclusive 'Cyber Security Center (CSC)' be formed under the NSC, which would be singularly responsible for policy formulation, budget allocation and nationwide implementation

**Cyber Crimes and Cyber Terrorism.** MHA should be the nodal agency for handling cyber terrorism. To handle cyber terrorism and cyber crime, a slew of measures will be needed, ranging from monitoring and surveillance, investigation, prosecution etc. The National Counter Terrorism Centre being set up should have a strong cyber component.

**Cyber Warfare.** There is a need to create a Directorate or Special Wing in the NSCS for this. It would oversee and coordinate both defensive and offensive cyber operations. Other aspects of Cyber Warfare to be looked into are:-

(a) **Raising of Cyber Command.** While cyber warfare is an ongoing activity during peace time there is a dire need to develop this capability for a warlike situation. Cyber warfare in a manner is Network Centric Warfare and will form an essential part of preparation of the battlefield in any future conflict. This will comprise not only the three Services but personnel from the DRDO and scientific and technological community.

(b) **Reserve of Young IT Professionals for Cyber Warfare.** There is a need to create and maintain a "surge capacity" for crisis or warlike situation. Young IT professionals constitute a vast resource base and a large number would be willing to loyally serve the nation when required. This resource must be capitalised by raising of cyber warfare reservists which could be embodied, when required.

**Capacity Building. Some of the measures are :-**

(a) There is need to place special emphasis on building adequate technical capabilities in cryptology, testing for malware in embedded systems, operating systems, fabrication of specialised chips for defence and intelligence functions, search engines, artificial intelligence, routers, etc. In the interim, all software and hardware manufactured by foreign Original Equipment Manufacturers (OEMs) needs to be tested for any security loopholes.

(b) Developing mobile software platforms including operating systems, anti viruses, root kits, malware, viruses, Trojans and other cyber weapons etc.

**Public Private Partnership.** Close cooperation between the Government and the Private Sector is necessary because much of the infrastructure and networks are in private hands. A joint working group was established in July 2012 with representatives from various ministries of the GoI and the Private Sector but it suffers from many problems like, the lack of a comprehensive road map with timelines and funding.

**International Cyber laws.** Adopting a proactive approach in the United Nations, including lobbying with like-minded nations in ensuring all encompassing international cyber laws and treaties are promulgated.

**Human Resource Development.** There is a need to introduce new courses, curriculum and academic institutions in the field of cyber security, ethical hacking, cryptology etc. to boost human resource in the field of cyber warfare.

**Synergy and Coordination.** There is a need for coordination, planning, understanding and synergy of efforts amongst all civil, military, intelligence, law enforcement and educational organisations responsible for cyber security, information assurance, cyber warfare and perception management.

**Research and Development.** There is a need to focus on :-[22]

(a) Functioning and Software design of social networks to ensure 'security and privacy', and emphasis on 'malware detection'.

(b) Develop reliable technology for protection of personal data in third party domain namely; social networks, cloud providers, outsourcing during various phases of its lifecycle; transmission, processing or storage.

(c) Develop mechanisms for ensuring digital rights and protecting privacy with assured empowerment of user to manage their data and avoid anonymous usage.

## Conclusion

The exponential growth of cyberspace is possibly the greatest development of the current Century. Cyberspace being the fifth common space, it is imperative that there be coordination, cooperation and uniformity among all agencies to safeguard it. There is no quick fix solution that can secure our cyber space. The solutions are sprinkled in strong policy and law enforcement, real-time information sharing, embracing technology and sensitising our cyber space users on cyber hygiene.

## Endnotes

1   'Internet in India 2014', jointly published by the Internet and Mobile Association of India (IAMAI) and IMRB International,

2   Facts revealed in a written reply to the Lok Sabha by Communications and IT Minister Ravishankar Prasad, on 03 Jul 14, 2014,

3   www.hackerslist.com accessed on 18 May 2015

4   www.techcrunch.com/fireeye-apt-30-southeastasia-india-report.html, Accessed on 18 May 2015

5   "Cyber Warfare: Pakistani Hackers Claim defacing over 2000 Indian Websites", Tribune, 02 Feb 2014.

6   Notification on National Cyber Security Policy 2013 by MoC&IT on 02 Jul 2013.

7   Extraordinary Gazette of India, Part II-section 1, No. 27 New Delhi, Friday, 09Jun 2013

8   Article by Sanjay Chhabra on 'India's National Cyber Security Policy and Organisation- Critical Assessment' in Naval War College journal

9   Op. Cit 7.

10  "Huawei spies for China, claims Ex-CIA Chief", Times of India, 19 Jul 2013.

11  http://www.idsa.in/monograph/Cybersecurity_csamuel.html accessed on 20 May 2015.

12  Ibid.

13  The "Fifty Second Report on Cyber Crime, Cyber Security, and Right to Privacy" issued by the 2013-2014 Parliamentary Standing Committee on Information Technology on 12 Feb 2014.

14  Ibid.

15  Op. Cit. 11.

16  C Henderson, 'The 2010 United States National Security Strategy and the Obama Doctrine of Necessary Force'.

17  United States Department of Defence, 'Quadrennial Defense Review -2014'.

18  Randy James (Time, US, 01 Jun 09) 'A Brief History of Cybercrime'.

19  US Def Sec Robert Gates memorandum (23 Jun 2003) 'Establishment of subordinate unified US Cyber Command under Strategic Command for Military Cyberspace Operations'.

20   US DoD Cyber Doctrine released on April 2015.

21   'The PLA as an Organisation' Volume v1.0 by Rand Publications 2002

22   'Cyber Security: Issues and Challenges", Article published in CSI
     Communications May 2015 by NJ Rao

‡

---

‡   **Lieutenant General Nitin Kumar Kohli, AVSM, VSM** was commissioned into the
    Corps of Signals on 17 Dec 1977. He commanded a Strike Corps Signal Regiment in
    the Western theatre, has been the Chief Signal Officer of a Corps and a Command; and
    was the Director General Manpower Planning at the Army HQ, prior to assuming the
    appointment of Signal Officer-in-Chief of the Indian Army on 01 Sep 2013.

# Cyber Non-State Actors: The Cyber Taliban[*]

## *Colonel Sanjeev Relia*

## Introduction

The world has got completely hooked to the information technology revolution. Computers, smart phones and internet have invaded into our lives to such an extent that our day to day functioning is now completely dependent on them. As we become more reliant on these technologies, we also expose ourselves to the dangers lurking around in the cyberspace. Cybercrime is one such danger. Millions of dollars are lost to cybercriminals every year. Yet cybercriminals are not the ones who pose the gravest of threats. It is the threat of presence of non-state actors in cyber domain that is worrying nations today. The very nature of cyberspace makes them a potent force that will play a pivotal role in any future cyberwar.

## Non-State Actors and Cyberwarfare

While warfighting is all about opposing armies battling it out and dominating each other in the air, sea and on land, non-state actors too have always played some role in all conflicts. The best example in the Indian subcontinent is the "Mukti Bahini" the Bengali resistance that fought against the Pakistan Army by the side of Indian Army during the Bangladesh Liberation War in 1971. Tasks from espionage to surveillance to physical combat all have been undertaken in the past by such armed non-state actors. But in the cyberspace this may not be the case. While in an armed conflict, it is the armed forces that play the most vital role, in a conflict through the cyberspace, non-state

---

[*] This article was first published in the *Journal of the United Service Institution of India*, Vol. CXLV, No. 599, January-March 2015.

actors may play a larger role than the armed forces would do in waging a war through this domain. This would be more so when the two nations are not in a state of armed conflict but hostilities do occur between them; e.g. India and Pakistan. We are not in a state of war, yet the relations between the two countries are not cordial. In such circumstances, non-state actors based in Pakistan and supported by Pakistan army/ government will play a crucial role in attacking our critical info-infrastructure through the cyber domain with the Pakistan army/government completely denying any involvement.

So who is a non-state actor in the cyberspace? They could be anyone from an ordinary citizen to a patriotic hacker to a cybercriminal to a cyber terrorist or even cyber militia. Past experiences of cyberattacks on Estonia in 2007 and Georgia in 2008 clearly show that the Russians, alleged of originating these attacks, completely denied any of its state machinery being involved in the attacks. Anonymity is a characteristic of cyber domain. The state machinery can, therefore, easily hide behind non state actors with little or no risk of attribution and deny any involvement in perpetrating devastating cyberattacks. In fact, the ease with which a cyber militia can operate and carryout cyberattacks, make them a better choice than establishing a full-fledged cyber wing as part of the armed forces.

## What is Cyber Militia?

A cyber militia can be defined as a group of volunteers who are willing and able to use cyberattacks or other forms of disruptive cyber actions in order to achieve a political goal.[1] They are men, not in uniform but motivated enough to be employed in covert government-orchestrated campaigns with the purpose to further the strategic political or military objective of the instigating state. It is said that China has established PLA Unit 61398 based at Shanghai staffed by thousands of computer professionals as "Cyber Troops" acting on direct orders of PLA.[2] Unit 61398 is supposed to be responsible for all major cyberattacks and cases of cyber espionage against the USA and other countries including India. China on the other hand completely denies even existence of any such unit, leave alone its involvement or connection of any other state machinery. But if reports in the western media are to be believed and also if Snowden revelations are correct, then China does have a potent group of non-state actors organised in the form of Unit 61398, acting completely under the control of PLA.

Employing cyber militia in place of regulars has tremendous advantages. Some of these are:-

(a) **Counterstrike.** Although employing non-state actors to carry out cyberattacks might raise suspicion in the international community, the lack of any hard evidence will protect the attacker of any political ramifications. Thus, the threat of a counterstrike is negligible. In 2007 while all evidence showed that the Distributed Denial of Service (DDoS) attacks on Estonia originated from Russia, Estonia or the NATO could not retaliate due to lack of attribution. While Russia completely denied any involvement, the execution may have been carried out by patriotic cyber militia on behest of the Russian government.

(b) **Cost Factor.** To raise a well organised cyber wing as part of the government or the defence forces would cost a lot of money as such a force will have to be funded and manned by uniformed personnel. By recruiting suitably motivated and technically competent non-state actors, the same task can be achieved at little or no cost. Small nation states today by sponsoring such cyber militia at negligible costs can threaten the critical infrastructure of much bigger and stronger nations.

(c) **Sponsor Cyberwar.** Non-state actors with the backing of state machinery can form unholy alliances, where state provides advanced capabilities in the form of money or actual intrusion tools to non-state actors who can then pass them on to another state or its non-state actors which wants to build cyberwar capability. As on date there are no international laws or treaties banning such actions. Hence, sponsoring a cyberwar through transfer of such technologies via non-state actors is perfectly legal, or atleast beyond reproach.

(d) **Freedom to Attack from Anywhere.** Non-state actors need not be based in the same country which is sponsoring them. Cyberspace knows no boundaries. Hence, the attack can be carried out with the same precision and impact with the attacker based in a third country. This makes the task of the attacking another nation even easier as attribution becomes even more difficult in such cases.

(e) **Laws of War do not Apply.** Even if an indisputable link is established between a non-state proxy and a nation-state, no laws of war apply to these cyber militias. This is because status of such non-state actors cannot legally be considered to be that of combatants. Also, in some cyberattacks, no physical damage may be caused by these cyberattacks; hence laws of armed conflict do not apply to them. Therefore, such non-state actors in the cyberspace may get away from being tried for war crimes despite the attacks having the same devastating impact as physical attacks.

Raising and employment of such cyber militia forces may have a flip side too. Just like there are no good or bad terrorists, similarly, there are no good or bad hackers. Armed with adequate knowledge and skills, the same attacker may turn against the state and threaten own infrastructure. They may even blackmail the government in order not to disclose sensitive details. Contracted cyber espionage agents might defect to the opposing nation if offered political asylum and cause damage like it happened in the case of Edward Snowden. However, the advantages of using such non-state actor outweigh the drawbacks. This is the reason that a number of nations are preferring employment of such forces instead of employing regular troops to attack the opponents through the cyber domain.

## What Threat Does India Face from Cyber Non-State Actors?

Anyone who deals with the cyberspace would know about Stuxnet and the crisis the computer worm created for Iranian nuclear programme in 2010. But many of us would not be aware that Stuxnet was detected in Indian hardware too. Based on a study of the spread of Stuxnet conducted by 'Symantec' an American computer security company, the most affected countries in the early days of the infection were Iran, Indonesia and India. As per a report released by Symantec in September 2010, 8.31 per cent computers in India were found infected with Stuxnet.[3] Stuxnet was designed to attack systems using certain specific software namely Windows Operating System, Siemens PCS 7, WinCC and STEP7 industrial software applications and one or more Siemens S7 PLCs. Only when presence of all software was detected by the worm, would Stuxnet be activated. If complete criteria were not met, the worm was programmed to destroy itself. This clearly indicates that Stuxnet was designed to target computers specifically associated with Supervisory Control and Data Acquisition (SCADA) systems as such software is found in SCADA/industrial control systems.

So how did the worm manage to reach well protected hardware in Iran and India and what damage was caused by it in India? Obviously no nation state was directly involved in perpetrating Stuxnet attack. The sophistication with which the worm's code was written and the lethality with which it carried out its task indicates that it was not a handiwork of some novice hacker. As no money or information was stolen by the exploits of the worm, it is unlikely that some motivated cyber criminals created and planted it to steal either money or information. That leaves only one option. The precision with which Stuxnet attacked SCADA systems indicate that it took a lot of planning and effort in implementation of the attack. Such a task could have been done either by cyber terrorists or non-state actors acting on behalf of some state. The same people who perpetrated the worm attack in Iran, also perhaps infected Indian systems also. While the damage caused by Stuxnet in Iran is well documented, unfortunately no survey is available in the public domain which could establish the nature of damage that may have been caused in India by it. Though some reports in the media indicate that INSAT- 4B a communication satellite launched by India in 2007 and which effectively went 'dud' in 2010 due to failure of its transponders affecting 70 per cent of Direct to Home services in India was a handiwork of Stuxnet.[4] The same has though not been confirmed by either ISRO or by Siemens whose software the satellite was using. Whether the satellite went 'dud' because of Stuxnet or not, the mere fact that such a deadly computer worm was able to penetrate unnoticed into control systems of our satellite network (if the Forbes report is to be believed), is an indication of the penetration capabilities of offensive cyber tools available today with rogue elements.

**Sabotaging the Critical Info-Infrastructure.** The above two incidences clearly indicate that networks and infrastructure in our country are vulnerable to cyberattacks, specifically by non-state actors acting on behalf of states like Pakistan or China. Sabotage is an integral part of Cyber Warfare. Malicious software and cyberattacks are ideal instruments of sabotage. This is especially applicable for sectors which provide direct services to consumers such as Telecom, Banking and Power sector. The above three sectors rely heavily on information and communication technology (ICT) and networking. As all of these three sectors provide consumer services, use of internet is also essential for all three sectors. While it is difficult to attack a standalone network or service, any infrastructure which is connected to the internet becomes more vulnerable to cyberattacks. Therefore these three sectors are specifically vulnerable to well-coordinated cyberattacks resulting in breakdown of their services. State sponsored non-state actors can not only target such critical info-infrastructure but other spheres of life which rely on ICT. As systems become more complex, the knowledge required to attack them also becomes more complex and arcane. Unless the attacker is backed up with full financial and knowledge support, sabotaging industrial control system will be a difficult task. Non-state actors are the only group of cyber adversaries who can achieve such a task with ease as they have all the necessary backing.

**Subversion.** Another activity which a non-state actor can undertake effectively through the cyberspace against our country is subversion. As per Thomas Rid, a British scholar and writer, information technology has enabled proliferation of subversive causes and ideas. Because of the cyberspace, subversion has become more cause driven, it is seeing higher levels of membership mobility and is now characterised by lower levels of organisational controls.[5] One common tool of all subversion activity is media, may it be print or visual media. The exponential rise and infinite reach of social media today has made it a perfect tool for subversive activities. The kind of influence social media has on the society has got our government thinking about the impact it can have on internal security of the Country. Today politicians, senior government officials and scholars can often be heard voicing their concern about the negative and subversive impact of social media. A very recent example of this was the exodus of the northeast students from Bangalore and other southern cities in August 2012. Despite appeals and assurances of safety by the Karnataka government, people from the northeastern parts of India working in cities of Karnataka continued to flee the state in hordes. Whatever were the actual reasons for the event, social media was blamed for the massive exodus.

Social media in particular and internet in general are mediums which a non-state actor can exploit for creating an adverse public opinion against the government of the day. Examples of this can be found in the way Arab Spring of 2011 was triggered. Social network, especially Facebook, offered a platform for planning and after action deliberations. The moderators of various Facebook groups that helped spark the unrest remained anonymous during most of the Arab Spring. Even the shutdown of the internet could not prevent the spread of political movement. The recent arrest of an ISIS Tweeter handler in Bengaluru shows the innovative ways a Jihadi organisation can make use of cyberspace. The IS militant group has made extensive use of social media for propaganda and recruitment, as well as for disseminating gory execution videos. If a banned jihadist rebel group based in Iraq and Syria can so well put to use the cyberspace, imagine how well a state sponsored organisation will be able to use it.

Listed above are just some of the ways a nation can employ non-state actors in the cyberspace. While sabotage, subversion and espionage would be the main motives behind employing cyber militia, there could be many other ways to use them in spreading terror in India using the cyberspace. Our armed forces and other governmental organisations have mastered the ways to counter state sponsored terrorism in J&K and the northeast; we will have to learn innovative methods for fighting actions perpetuated through the cyber domain. Time has come to recognise the potential of non-state actors in the cyberspace and take countermeasures against their likely method of operations.

## Conclusion

Non-state actors wield more influence and pose greater national security risks in the cyber domain than they do on land, sea and air. With low barriers to entry and the ease with which technology today is available, a state can achieve its nefarious goals in the cyber domain by proxy non-state actors who can be as effective as a nation state in undertaking precision cyberattacks. It is time that the government took a serious view of this and addressed the issue of cyber conflict with non-state adversaries. It is a must to establish a secure and resilient cyberspace in the Country.

## Endnotes

1 Ottis, R., "Proactive Defense Tactics Against On-Line Cyber Militia," in the proceedings of the 9th European Conference on Information Warfare and Security (ECIW 2010), Thessaloniki, Greece, Jul. 2010.

2 China hacking charges: the Chinese army's Unit 61398 as available at http://www.telegraph.co.uk/news/worldnews/asia/china/10842093/China-hacking-charges-the-Chinese-armys-Unit-61398.html

3 "W32.Stuxnet". Symantec. 17 September 2010 available at http://www. symantec. com /security_respon se/writeup.jsp?docid=2010-071400-3123-99 viewed on 20 Dec 2014

4 Did The Stuxnet Worm Kill India's INSAT-4B Satellite? As available at http://www.forbes.com/sites/firewall/2010/09/29/did-the-stuxnet-worm-kill-indias-insat-4b-satellite/ viewed on 20 Dec 2014.

5 Cyber War Will Not Take Place by Thomas Rid P115

† **Colonel Sanjeev Relia** was commissioned into the Corps of Signals on 20 Dec 1986. Presently, he is a Senior Research Fellow at the Centre for Strategic Studies and Simulation, United Service Institution of India, New Delhi.

# Living With Cyber Surveillance and Espionage*

## Colonel Sanjeev Relia

## Introduction

Surveillance and espionage have existed for time immemorial. While they were always a part of any military campaign and study, private lives of ordinary citizens were generally not much affected by such activities. Things are not the same anymore. With the internet invading into our lives like never before, we today live under constant surveillance of multiple agencies like the government, your employer and perhaps your friends and neighbours. While a lot has been written and spoken about surveillance using land, sea, air and space, not much is heard about the surveillance using the fifth domain – the cyber space.

In the age of internet, where information travels at the speed of light and events can be created in a matter of microseconds, privacy and safeguarding one's personal information has become a challenge. Securing private information was so much simpler when there was no internet. Today, Google perhaps knows more about what you do when, than you yourself know. The recent one minute video clip created by the social media Facebook for all its users clearly indicates that personal information is not as personal as we consider it to be and that someone is constantly watching over you.

## What are Cyber Surveillance and Cyber Espionage?

Surveillance is defined as monitoring of the behaviour, activities or other changing information, usually of people for the purpose of influencing, managing, directing, or protecting them.[1] Technology has always played a very

* This article was first published in the *Journal of the United Service Institution of India*, Vol. CXLIV, No. 596, April-June 2014.

important part in such monitoring activities. In the realms of surveillance, the impact of networking technologies has been phenomenal which has given rise to a new model of surveillance called Cyber Surveillance.

Cyber surveillance is monitoring of computer activity, data stored on a hard drive, or being transferred over computer networks such as the Internet. Monitoring is often done clandestinely and may be done by or at the behest of governments, by corporations, criminal organisations, or individuals. It may or may not be legal and may or may not require authorisation from a court or other independent agency.[2] In 2013, Edward Snowden, a former employee of the CIA and then a contractor working for the National Security Agency (NSA), revealed the scale of America's secret mass cyber surveillance programme at the transnational level codenamed "PRISM". (Interestingly Snowden is said to have learnt his hacking skills in an ethical hacking institute in India). Snowden's leaked documents uncovered the existence of a large number of global surveillance programmes, many of them run officially by the USA. While some called Snowden a hero for exposing the clandestine cyber activities of the Obama administration, there are some who also referred to him as a traitor. His disclosures nonetheless have fuelled global debates over mass cyber surveillance, government secrecy and the balance between national security and information privacy.

Cyber espionage on the other hand is the act or practice of obtaining secrets without the permission of the holder of the information (personal, sensitive, proprietary or of classified nature), from individuals, competitors, rivals, groups, governments and enemies for personal, economic, political or military advantage using methods on the Internet, networks or individual computers through the use of cracking techniques and malicious software.

## How is Cyber Surveillance/ Espionage Done?

The global surveillance industry is estimated to be growing at over 5 billion US dollar a year. Capabilities of surveillance technology have grown vastly in the past decade. Today, cyber surveillance technology ranges from malware which infects a target computer to record every keystroke, to systems for tapping undersea fibre-optic cables in order to monitor the communications of entire population. Some of the common surveillance techniques being used to gather information from the cyber domain are covered in the succeeding paras.

**Network Surveillance.** Majority of computer surveillance involves the monitoring of data and traffic while it is moving on the network. This includes both the Internet as well as stand alone discreet networks. The USA leads the pack of nations who indulge in such activities. While all phone calls and broadband internet traffic are required to be made available for real time monitoring under the Communication Assistance for Law Enforcement Act in the USA, international traffic moving on the internet too is susceptible to monitoring. Figure 1 below which was part of the secret PRISM presentation clearly indicates why any communication originated in India for a recipient in Africa or Europe can be easily tracked and monitored in the USA.[3]

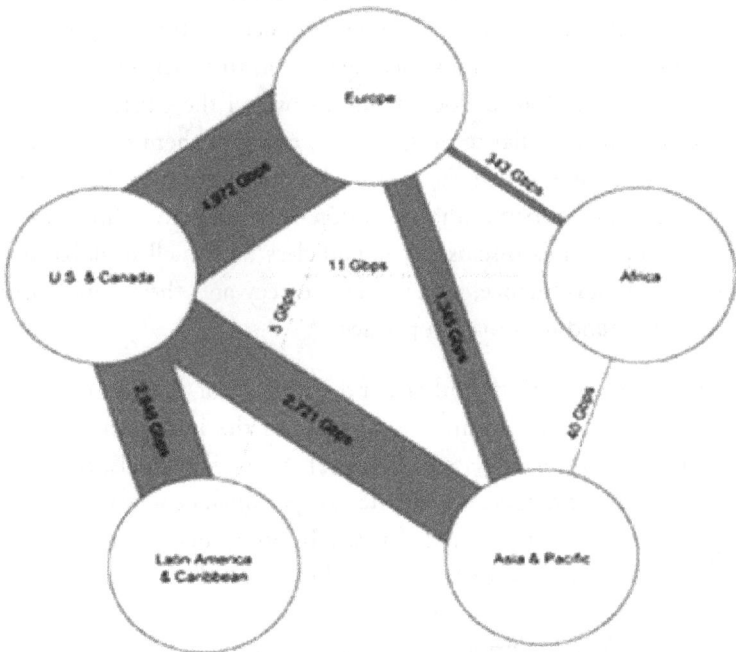

Figure 1 : International Internet Regional Bandwidth Capacity in 2011

As is clearly evident from the above diagram, bulk of the internet traffic flows via USA. Also, over eighty per cent of the servers and cloud providers are located in the USA. PRISM findings indicated that the internet service provider companies in the US whom the world trusted with their most private data were handing over the data to the NSA as if they too were a part of the entire clandestine operation.

**Packet Capture.** Packet capture or packet sniffing is monitoring of data traffic on a computer network. Data over the internet is transmitted by breaking it into small chunks called "packets", which are routed through a network of computers. At destination, they are reassembled back into original message. Packet Capture Applications intercept these packets as they are travelling through the network, in order to examine their contents using analysis tools thereby deriving information out of them. As there is far too much of data travelling on the internet, automated Internet surveillance computers sift through the vast amount of intercepted Internet traffic to filter out information based on keywords or phrases, visiting certain types of websites, or communicating via e-mail or chat with a certain individual or group. Deep packet inspection (DPI) is the leading method of such surveillance. DPI technologies are capable of analysing the actual content of the traffic that is flowing. DPI allows network operators to scan the payload of IP packets as well as the header. This technique is often employed by law enforcement agencies and security forces trying to identify cyber criminals and cyber terrorists over the internet.

**Malicious Software.** Use of computer viruses, worms and Trojans is an effective method to examine or steal data stored on a computer's hard drive, as well as to monitor activities of a person using the computer. A surveillance programme maliciously installed on a computer can search the contents of the hard drive for data, monitor computer use, collect passwords, and report back activities in real-time to its controller through the Internet connection. GhostNet a Trojan used by the Chinese in 2008-09 is an example of cyber surveillance of the Tibetans community in general and Dalai Lama in particular, using the internet. This Trojan allowed attackers to gain complete, real-time control of the infected machines and diverted the data to its controllers in island of Hainan. How deep routed is the Chinese cyber espionage set-up can be made out from Information Warfare Monitor (IWM) investigation detailed report of GhostNet, an extract of which is under :-[4]

*"During the course of our research, we were informed of the following incident. A member of Drewla, a young woman, decided to return to her family village in Tibet after working for two years for Drewla. She was arrested at the Nepalese-Tibetan border and taken to a detention facility, where she was held incommunicado for two months. She was interrogated by Chinese intelligence personnel about her employment in Dharamsala. She denied having been politically active and insisted that she had gone to*

*Dharamsala for studies. In response to this, the intelligence officers pulled out a dossier on her activities and presented her with full transcripts of her Internet chats over the years. They indicated that they were fully aware of, and were monitoring, the Drewla outreach initiative and that her colleagues were not welcome to return to Tibet. They then released her and she returned to her village."*

**Social Network Analysis.** Social media technologies such as Facebook and Tweeter can be used by companies, marketers, and governments to collect significant amounts of data about individual users. Aim of this form of cyber surveillance is to create maps of social networks based on data from social networking sites as well as from traffic analysis information from phone call records. These social network "maps" are then data mined to extract useful information such as personal interests, friendships and affiliations, wants, beliefs, thoughts, and activities.

**Hardware Monitoring.** Today techniques exist where network or computer transmissions can be monitored using the hardware installed in the system. Some of these techniques are :–

(a) Monitoring by detecting the radiations emitted by the Cathode Ray Tube monitor. In the USA, surveillance using spurious transmissions being emitted by hardware is termed as TEMPEST.

(b) Using the transmissions between a computer and a presentation device such as a projection system.

(c) Picking up the noise of the key board clicking. Research shows that each key has a distinct noise which can be picked up using an audio surveillance device and the message deciphered.

(d) Use of the audio speakers connected to the system for picking up and transmitting information from a computer.

(e) Radio Pathways. Tiny trans-receivers are built into Universal Serial Bus (USB) plugs and inserted into target computers which then communicate with a hidden relay station up to 12 kms away. This method is most effectively used for machines isolated from the internet.[5]

**Supply Chain Vulnerability.** A supply chain attack is an attack through subversion of hardware or software supply chain. A product, typically a device that performs encryption or secures transactions, is tampered with during manufacture or while it is still in the supply chain by persons with physical access. The tampering may, for example, install a root-kit or hardware-based spying components. The aim is to first gather information from the place where this hardware or software is installed and then to execute a cyber attack. Unless the user has facilities of test labs where such equipment and the software can be checked for spyware, a supply chain malware can never be detected. Countries like India where most of the public as well military hardware and software are imported remain vulnerable to supply chain surveillance and attacks. Although National Cyber Security Policy 2013 does talk of undertaking R&D programmes by the government for addressing all aspects including development of trustworthy systems, their testing, development and maintenance throughout their life cycle, is still a far fetched dream. It is unlikely that we will ever be able to sanitise hundred per cent hardware and software being used in the Country. But even if we are able to sanitise the critical components being installed in the national info-infrastructure, we would be able to save ourselves from loss of critical information and Stuxnet kind of attacks.

## Legal Cover to Surveillance in India

The Economic Times in December 2013 reported that the Government of India will shortly launch 'Netra', the defence ministry's internet spy system that will be capable of detecting words like 'attack', 'bomb', 'blast' or 'kill' in a matter of seconds from reams of tweets, status updates, e-mails, instant messaging transcripts, internet calls, blogs and forums. The system will also be able to capture any dubious voice traffic passing through software such as Skype or Google Talk.[6] The 'Netra' internet surveillance system has reportedly been developed by Centre for Artificial Intelligence & Robotics (CAIR), a laboratory under the DRDO.

So, does this mean that the Indian Government too has officially announced that all internet traffic in the country is liable to monitoring like the US Government did in PRISM? The Information Technology (IT) Act of 2000 and IT Act Amendment 2008 does give the power to the Government to carry out monitoring of the internet. Section 69(B) confers on the Central Government power to appoint any agency to monitor and collect traffic data or information generated, transmitted, received, or stored in any computer resource in order to enhance its cyber security and for identification, analysis, and prevention of intrusion or spread of computer contaminant in the country. Under this section, any government official or policeman can (or perhaps is already doing) listen in to all your phone calls, read your chats, SMSs and e-mails, and monitor the websites you visit. No search warrant from a magistrate is required to do so by them.

While the Act is a good tool to control cyber crime and cyber terrorism, Section 69(B) of the IT Act Amendment 2008 gives unrestrictive powers of the Government and law enforcement agencies in the Country which can be used to snoop upon unsuspecting citizens. Today the internet is a central element of the info-infrastructure of the information society and a global facility available to the public. The global and the open nature of the internet is a driving force in accelerating progress towards development in its various forms. It is, therefore, important to maintain an open environment that supports free flow of information across the globe and hence, it is essential that surveillance of such a resource, especially by nation states is dealt with caution.

Some governments across the globe argue that internet surveillance is necessary to ensure national security. As per them, keeping an eye on the data flowing over networks is a key to keep the nation safe. The unprecedented Chinese Government's programme of censorship of its people is an example of such a policy. The surveillance and content control system, launched in November 2000 by Peoples Republic of China, became known as the Great Firewall of China, where every bit of information flowing on the internet is kept under a watch by the Communist Government. There are also nations who feel that any such clandestine surveillance undertaken is a violation of human rights especially freedom of information. The stand of the Indian government is not too clear on this aspect. However, it is the duty of any democratically elected Government to appreciate that every law abiding citizen has the right to have a private life, a life which is not fully under constant surveillance of any state machinery.

## Impact of Cyber Surveillance and Espionage on the Society

The Justice Department of the United States of America on 19 May 2014 filed unprecedented criminal charges against the members of the Chinese military, accusing them of economic espionage by hacking into the computers of US companies involved in nuclear energy, steel manufacturing and solar energy. Chinese government strongly rebuked the US over its claims of cyber-spying saying they were based on "fabricated facts" and would jeopardise US-China relations. Whether the charges are true or not, the fact of the matter is that cyber surveillance and espionage today has reached a level where it has started to impact bilateral relations between two strong nations.

The economic and business world suffers enormously form malicious cyber activities. While it will be difficult to gauge total cost to societies of cyber surveillance and espionage, but as a rough estimate as per a 2011 research, the upper limit of the cost of cyber espionage and crimes is somewhere between 0.5 per cent and one per cent of the national income. Also, not only does cyber espionage contributes towards high financial losses, it also has intangible loss component associated with it such as:

(a) The loss of intellectual property.

(b) The loss of sensitive business information (such as negotiating strategies), including possible stock market manipulation.

(c) Opportunity costs, including service disruptions, reduced trust online, the spending required in restoring any "lead" from military technology lost to hacking, and the realignment of economic activity as jobs flow out of "hacked' companies.

(d) The additional cost of securing networks and expenditures to recover from cyber-attacks.

(e) Reputational damage to the hacked company.

Here is an example of how colossal damages can be inflicted to even a powerful nation like the USA through cyber espionage. As per a news report, a cyber espionage operation by China seven years ago resulted in stealing sensitive technology and aircraft secrets that have now been incorporated into the latest version of China's new J-20 stealth fighter jet. The Chinese cyber spying against the Lockheed Martin F-35 Lightning II took place in 2007. Stolen data was obtained by a Chinese military unit called Technical Reconnaissance Bureau in the Chengdu province. The F-35 data theft was confirmed by the USA after some photographs of the J-20 were published on Chinese websites showing a newer version of the aircraft.[7] The loss of critical design information of the F-35 was part of widening Chinese campaign of espionage against the US defence contractors and government agencies.

## Conclusion

Invasion of the internet into our daily lives is a relatively recent phenomenon. Life is getting more and more dependent on the cyber world. Today the fear that surveillance can actually become so extensive as to threaten an individual's healthy moral development is reasonable. Hence, the society needs to guard against it.

Most of the world is inadequately prepared for defending against these new types of surveillance and espionage techniques which have emerged in the last two decades. Governments, businesses, organisations, individual owners and users of cyberspace must assume responsibility for and take steps to enhance the security of the information technologies against such malicious cyber activities. Then only will this resource contribute towards positive growth of the society.

### Endnotes

1. Lyon, David. 2007. Surveillance Studies: An Overview. Cambridge: Polity Press.

2. As available at http://en.wikipedia.org/wiki/Computer_and_network_surveillance accessed on 30 Apr 2014.

3. So just what exactly is NSA's PRISM by Rick Falkvinge as available at http://falkvinge.net/2013/06/08/so-just-exactly-what-is-nsas-prism-more-than-reprehensibly-evil/ accessed on 30 Apr 2014.

4. Tracking GhostNet: Investigating a Cyber Espionage Network, Information Warfare Monitor, 29 March 2009, available at http://www.nartv.org/mirror/ghostnet. pdf. Accessed on 30 Apr 2014.

5. NSA uses secret radio pathways to spy on offline PC's, Times of India, 16 January 2014

6. Government to launch 'Netra' for internet surveillance, Kalyan Parbat, ET Bureau Dec 16, 2013 as available at http://articles.economictimes. indiatimes. com/ 2013-12-16/news/45256400_1_security-agencies. Accessed on 30 Apr 2014.

7. Top Gun takeover: Stolen F-35 secrets showing up in China's stealth fighter, by Bill Gertz, Washington Times March 13, 2014.

† @Colonel Sanjeev Relia was commissioned into the Corps of Signals on 20 Dec 1986. Presently, he is a Senior Research Fellow at the Centre for Strategic Studies and Simulation, United Service Institution of India, New Delhi.

# Cyber Warfare and the Laws of War[*†]

## Wing Commander UC Jha, PhD (Retd)

## Introduction

Today, computers control much of our civilian as well as military infrastructure, including communications, power systems, banking and healthcare. The Internet provides nearly universal interconnectivity of computer networks, with no distinction being made between civilian and military uses. It has the capacity to carry data across the countries, continents and oceans of the world. The Internet has expanded rapidly at a global scale and has been the most powerful technological revolution known in the history of mankind. In just 18 months, between December 2010 and June 2012, the number of individuals actively using the Internet increased from an estimated 36 million to more than 2.4 billion. There has also been a phenomenal growth in military reliance on computer systems. This has introduced a "fifth" domain in which wars may be fought, besides the conventional domains of land, sea, air and space. Given the increasing reliance on information systems in general and access to the Internet in particular, critical military and civil infrastructure is growing more and more vulnerable to cyber attacks. Some have even likened the potential of cyber weapons to inflict damage to that of nuclear weapons.

Cyber warfare, unlike nuclear warfare, is not just the province of the industrial nation-state. Terrorist groups, whether state-sponsored or independent, domestic or international, as well as organised crime syndicates

* This is the edited text of a Talk delivered by Wing Commander UC Jha, PhD (Retd) on 26 Feb 2014 at USI, with Lieutenant General SP Kochhar, AVSM**, SM, VSM (Retd) in the Chair.

† This article was first published in the *Journal of the United Service Institution of India*, Vol. CXLIV, No. 595, January-March 2014.

and individuals, are equipped with cyber technologies with which they can launch cyber attacks. The potential of cyber capabilities to cause serious harm to an adversary is no longer theoretical. During the Cold War, the Central Intelligence Agency (CIA) allegedly gained unauthorised access to a Soviet computer to install a malicious code, called a logic bomb, which the CIA subsequently used to destroy a Soviet natural gas pipeline in 1982. An expertly conducted cyber attack could destroy a nation's economy and deprive much of its population of basic services, including electricity, water, sanitation, and health. Cyber attacks and cyber warfare undoubtedly present new and difficult legal problems.

## Cyber Warfare

Cyber activities can span from cyber crime to cyber espionage to cyber terrorism and all the way to cyber attacks and cyber warfare. The term cyber warfare refers to warfare conducted in cyberspace through cyber means and methods. Warfare is commonly understood as the conduct of military hostilities in situations of armed conflict. Cyber attacks comprise efforts to alter, disrupt or destroy computer systems or networks or the information or programs on them. They may vary in terms of target (military versus civilian, public versus private), effect (minor versus major, direct versus indirect), and duration (temporary versus long-term).

Cyberspace is a global domain consisting of the interdependent networks of information technology infrastructures and resident data, including the Internet, telecommunication networks, computer systems, and embedded processors and controllers. It is the only domain which is entirely man-made. It is created, maintained, owned and operated collectively by public and private stakeholders across the globe and changes constantly in response to technological innovation. It is not subject to geopolitical or natural boundaries, and is readily accessible to governments, non-state organisations, private enterprises and individuals alike.

## Cyber Warfare and the Use of Force

The most important source of the body of law, i.e., jus ad bellum, which governs the "use of force" by States in their international relations, is the UN Charter.[1] Article 2(4) of the Charter provides that all Members shall refrain in their international relations from the threat or use of force against the territorial integrity or political independence of any state, or in any other manner inconsistent with the purposes of the United Nations. The "use of force" constitutes an internationally wrongful act entailing the international responsibility of the State, and also allows the victim State to take counter-measures against the perpetrator. The Charter allows for two exceptions to this prohibition; the right to self-defence in case of an armed attack as well as the use of force authorised by the UN Security Council. A state-sponsored cyber operation would qualify as a use of force against another State and may also trigger an international armed conflict. A cyber operation amounting to an armed attack would permit the attacked state to exercise its inherent right to self-defence. However, a cyber operation that merely causes inconvenience or irritation would not qualify as use of force.

## Action in Self-defence

A State that is the target of a cyber operation which is equivalent to an armed attack may exercise its inherent right of self-defence. Whether a cyber operation qualifies as an armed attack depends on its scale and effects. It would be immaterial whether the cyber attack is against a military target or civilian objects. It would be considered an attack against the State. However, a cyber attack on any civilian infrastructure cannot be considered an armed attack. Though there are no agreements on what is critical infrastructure, the UN General Assembly (A/RES/58/199 of 23 December 2003) has recognised that "each country will determine its own critical information infrastructure". The UK, the US and Australia include agriculture, food, water, public health, emergency services, government, defence industrial base, information and telecommunication, energy, transportation (aviation, maritime and surface), banking and finance, chemicals and hazardous materials, and postal system among critical infrastructure. However, this is not conclusive; any system related to a State's economic prosperity, public safety and national defence would constitute a critical infrastructure.

The action of a State in self-defence against a cyber attack must meet the requirements of necessity, proportionality and immediacy. This means the use of force is the last resort, only if the matter cannot be settled by peaceful means. Further, there is an obligation to identify the author and verify that the cyber attack was not accidental. The same rules apply to cyber capabilities as to traditional kinetic weapons. A State could also resort to anticipatory self-defence against an imminent attack through conventional means.

## Cyber Attack: Legal Obligations

The 1977 Additional Protocol I (AP-I) to the Geneva Convention illustrates the principle of distinction to protect civilians during armed conflict. Under this principle, parties to an armed conflict must always distinguish between civilians and civilian objects on the one hand, and combatants and military targets on the other. Under AP-I civilians and civilian objects cannot be targets of attack. The treaty bars belligerents from rendering useless those objects that are indispensable to the survival of the civilian population, such as foodstuff, agricultural crops, livestock, drinking water installations and supplies, and irrigation works. Further, the States must never use weapons that are incapable of distinguishing between civilian and military targets. In the conduct of military operations, belligerents have the duty (i) to exercise constant care to minimise the loss of civilian lives and damage to civilian objects; (ii) to protect the natural environment and protect works and installations containing dangerous forces, such as dams and nuclear power plants; and (iii) not to undertake attacks that have the primary purpose of spreading terror among the civilian population.

In planning a cyber attack, military commanders must comply with the principle of distinction as well as proportionality. There are a few situations where the principle of distinction could be easily applied to cyber attacks, such as when the target is a military air traffic control system and the attack causes a troop transport to crash. If properly executed, the result of the cyber strike would be the same as a conventional bombing. However, often it may be nearly impossible to distinguish between combatants, civilians directly participating in hostilities, civilians engaged in a continuous combat function, and protected civilians in the context of cyber attacks. The obligation of legal review of new weapons, means or methods of warfare are contained in Article 36 of AP-I. These obligations are a part of customary international law[2] and applicable to cyber weapons too.

## Non-international Armed Conflict

Sophisticated non-State actors can also launch severe cyber attacks against the government, affecting the economy and communication. Non-State actors committing cyber crime and economic cyber espionage do pose serious threats; however, till date there are no reports of highly devastating cyber attacks launched by non-State actors against a State. With technical advancement and the proliferation of malware tools; or with support from a technically advanced State, the possibility of non-State actors carrying out sophisticated cyber operations cannot be dismissed. For instance, hijacking of drones by insurgents in future conflicts cannot be ruled out. Some students in Texas recently managed to take a ship completely off-course (off Italy) by interfering with GPS signals.[3]

## The Status of Cyber Warriors

Cyber warriors could be classified into four major categories; namely, combatants, contractors and civilian employees of the armed forces, levee en masse, and civilians. Cyber operations are generally carried out by highly specialised personnel. To the extent that they are members of the armed forces of a belligerent State, their status, rights and obligations are no different from those of traditional combatants. According to the laws of war, the armed forces of a belligerent State comprise all organised armed forces, groups and units which are under a command responsible to that State for the conduct of its subordinates. This broad and functional concept of armed forces includes essentially all armed actors belonging to a belligerent State and showing a sufficient degree of military organisation.

In the last two decades, belligerent States have increasingly employed private contractors and civilian employees to perform a variety of functions traditionally performed by military personnel. This includes the support, preparation and conduct of cyber operations. As long as such personnel assume functions not amounting to direct participation in hostilities, they remain civilians. In case they are formally embedded in the armed forces in an armed conflict, they would be de facto irregular members of the armed forces and entitled to prisoner of war status in case of capture.

The concept of a levee en masse is that during the initial invasion, the civilian population of unoccupied territory can spontaneously 'take up arms against the invading army' in order to forestall an occupation. The mobilisation of a levee en masse is patriotic zeal coupled with the initiative of the citizen-soldier under emergency until the enemy has been defeated or repelled. The law of war recognises the concept and protects those who participate in a war and 'carry their arms openly' by granting them combatant status under the Geneva Convention of 1949. While this category of persons has become ever less relevant in traditional warfare, it may well come to be of practical importance in cyber warfare. In cyber warfare, territory is neither invaded nor occupied, which may significantly prolong the period during which a levee en masse can operate. Also, cyber space provides an ideal environment for the instigation and non-hierarchical coordination of spontaneous, collective and unorganised cyber defence action by great numbers of "hacktivists". The only problem foreseen in this case is that in the context of cyber space, how would the requirement of "carrying their arms openly" be interpreted?

Under the laws of war, civilian means all persons who are neither members of the armed forces of a State or non-State party to an armed conflict, nor participants in a levee en masse. As civilians, they are entitled to protection against the dangers arising from military operations and against attack. In cyber warfare, this category is likely to include most non-State hackers not belonging to the armed forces. If and for such time as their operations amount to direct participation in hostilities, civilians lose their protection and may be directly attacked as if they were combatants. They do not benefit from immunity from prosecution for lawful acts of war and, therefore, can be punished by their captor for any violation of national law.

## Tallinn Manual

One of the problems with cyber warfare is the lack of uniformity in concepts, definitions, rules, policy and law. In many instances, not only is uniformity lacking, but there is simply a void. As a result, there is no general international consensus on how to treat cyber warfare. Attempts have been made, however, to create concepts, definitions, rules, policy and law regarding cyber warfare. The Tallinn Manual on the International Law Applicable to Cyber Warfare has been prepared by experts working with the Cooperative Cyber Defense Center of Excellence (CCDCOE), an institute based in Tallinn, Estonia, that assists NATO with technical and legal issues associated with cyber warfare. The Manual, released in 2013, is particularly concerned with jus in bello (the law of war) and jus ad bellum (the set of rules to be consulted before engaging in war) and does not deal with cyber crime in general or cyber terrorism. It is intended as a reference for legal advisers for government agencies.

The Manual consists of 95 rules reflecting customary international law and has been adopted unanimously by an International Group of Experts. It defines a cyber attack as "a cyber operation, whether offensive or defensive that is reasonably expected to cause injury or death to persons or damage or destruction to objects." The definition makes it clear that a cyber attack is an act of violence either against a person or object and that the focus is on the consequences and not the initiating act itself. Thus, the consequences of a cyber attack must generate some violence to some person or property. Therefore, it is not the act itself, but rather the subsequent consequences thereof that matter. The Manual has been criticised for being "an exercise of academic debate, restating what has been the practice, but failing to address the central issues raised by the emerging technical landscape." Developing international law for cyber warfare is a complex challenge and will take many nations coming to an agreement over a substantial period of years.[4] It is not possible to base it on some basic and fundamental concepts, definitions and rules created by some influential countries, especially when it relates to the safety and security of a State.

## The Future

Governments as well as industries have established both formal and informal mechanisms for countering rapidly increasing cyber threats and operations. More than 100 militaries in the world have dedicated cyber-attackers and defenders and have built some kind of cyber military units. These include the establishment of the US Cyber Command, China's People's Liberation Army General Staff Department's 3rd Department, Iranian Sun-Army and Cyber Army, Israel's Unit 8200, and the Russian Federal Security Service's 16th Directorate. India may soon have an independent Cyber Command to protect the nation's cyber domain and vital infrastructure.

There are varying opinions on how to tackle the issue of cyber warfare. While few are in favour of an international convention, others have opposed efforts to create a new treaty and have argued that the current laws of war can be applied to cyber warfare by analogy. It is clear, however, that States must develop a cyber warfare doctrine (CWD) to regulate the use of cyber weapons in war. India has been a major target of cyber attacks and the frequency and intensity of such attacks is increasing. There have been numerous incidents of sensitive government and military computers being attacked by unknown entities and information being stolen. India should be prepared for a cyber attack and stand ready to launch a counter-offensive. We must find answers to issues such as what activities must be undertaken in the case of a cyber attack against our nuclear power plants; what would be the appropriate response in the case of such an attack; and the attack threshold that would constitute an act of war. The CWD must be based on our current legal doctrine and precedents.

# Conclusion

Compared to the weapons that threatened States in the past, modern technology has made the tools of cyber warfare cheap, readily available and easily obtainable. Legal norms are emerging in cyber warfare, but many questions about what is legal and what is not in this "fifth domain of warfare" need to be answered. The implication of future cyber warfare is uncertain. Cyber weapons, while targeting military objectives, may also attack civilian objects such as railways, air traffic, hospitals, and power plants, causing massive collateral damage and civilian casualties. In addition to the legal regulation of cyber warfare, cyber espionage, theft of intellectual property, and a wide variety of criminal activities in cyberspace pose real and serious threats to all States, as well as the corporate world and private individuals. An adequate response to issues related with cyber crimes requires national and international measures. It is important that States be aware not only of their legal duty to examine whether new weapons and methods employed in cyber warfare would be compatible with their obligations under existing laws of war, but also of their moral responsibility towards generations to come.

# Endnotes

1.  Jus ad bellum is the Latin term for the law governing the resort to force i.e. when a State may use force within the constraints of the UN Charter framework and traditional legal principles. The modern jus ad bellum has its origins in the 1919 Covenant of the League of Nations and the UN Charter.

2.  The customary international law requirement for legal review of a weapon to ensure its use will be lawful in conflict stems from the 1868 St Petersburg Declaration, the 1899 Hague Declaration Concerning Asphyxiating Gases, the 1899 Hague Declaration Concerning Expanding Bullets and the 1907 Hague Convention IV Respecting the Laws and Customs of War on Land. These international instruments address the issue of whether a weapon causes superfluous injury in violation of the laws and customs of warfare. Additionally, in 1996, the International Court of Justice confirmed this customary international law status in its Nuclear Weapons Opinion.

3.  Civilian GPS is vulnerable to being spoofed, 14 August 2013. Available at: http://www.technologyreview.com/news/517686/spoofers-use-fake-gps-signals-to-knock-a-yacht-off-course/, accessed 23 January 2014.

4. James E McGhee, Cyber Redux: The Schmitt Analysis, Tallinn Manual and US Cyber Policy, Vol. 2 (1), Journal of Law & Cyber Warfare, Spring 2013, pp. 64-103, at p. 93.

‡ Wing Commander U C Jha, PhD (Retd) served in the IAF for 24 years and took premature retirement in 2001. He is an independent researcher in the field of environmental law, human rights, international humanitarian law and military law. He has authored a number of books, the latest one being, "Drone Wars: Ethical, Legal and Strategic Aspects", a USI publication.

# Understanding Cyber Weapons[*]

## *Colonel Sanjeev Relia*

## Introduction

A weapon is any device used in order to inflict damage or harm to living beings, structures, or systems. Weapons are used to increase the efficacy and efficiency of activities such as hunting, crime, law enforcement, self-defence, and warfare.[1] In a broader context, weapons may be designed to include anything used to gain a strategic, material or psychological advantage over an adversary. Simple as well as complex products can be used both for peaceful purposes and as arms or weapons depending on the intentions of the user. So is the case with Information Technology tools. While the internet was never designed to wage a war, cyber space today is on the threshold of being the fifth domain of warfare.

A lot is being spoken about attacks in the cyber space and cyber weapons. Nations like the USA, Russia and China have gone ahead and raised cyber units and even cyber commands. Although, we in India lag behind a little in this facet of warfare, however measures have been initiated to bridge the gap. The defence minister of India, Shri AK Antony in May 2013 announced that India too will soon form a cyber command to handle the online threats being faced by the country. He further said that we have already got a mechanism for cyber security but we are augmenting it further and the forces are finalising a proposal for a cyber command.[2] Does this mean that we have entered into an era where our defence forces will be fighting in the cyber domain? Does it also imply that we will need to create new kind of weapon systems to ensure territorial integrity of our nation and that we will have to create new units of

---

[*] This article was first published in the *Journal of the United Service Institution of India*, Vol. CXLIII, No. 594, October-December 2013.

computer geeks in uniform who will control cyber weapons and employ them when the need arises?

## What are Cyber Weapons?

When a weapon is spoken or written about, a distinct perfect shape of the weapon system emerges in the mind. Whatever be the weapon system, the IHS Jane's Defence Weekly would clearly spell out the destructive powers of each of the system with a few glossy pictures leaving little doubt in the mind what the weapon platform or system can do. But as the concept of cyber weapon is abstract, these weapons find no such mentions in any of such glossy journals. Besides, what can be made out of a printout of a software programme claiming to be a deadly cyber worm capable of destroying the command and control system of an Air Defence Network? A trained military mind finds it difficult to comprehend how an innocent looking laptop or desktop could cause havoc in the battlefield and perhaps may cripple the critical infrastructure of a nation. Therefore, there is a need for the men in uniform to critically analyse the concept and capabilities of Cyber Weapons.

The definition of a kinetic weapon fails to capture the essence of what are generally regarded as cyber weapons. This is because most of the malicious computer codes, may it be Virus, a Trojan or a Worm that would fall within the parameters of a cyber weapon are designed to have an indirect kinetic outcome which may, or may not, result in inflicting damage or harm to living beings, structures, or systems. To define a cyber weapon in the specific context of conflicts, it is necessary to first differentiate a cyber weapon from a malware, typically used for criminal purposes or an information tool used to perform espionage in the cyber space. To reach a definition of cyber weapon, it is therefore necessary to focus on three essential elements. As Thomas Rid explains it, a computer code as a cyber weapon has to first "weaponise" the target system in order to turn itself into a weapon.[3]

(a) **The Context.** It must be the typical context of a cyber warfare act. This concept may be defined as a conflict among actors, both national and non-national, characterised by the use of technological information systems, with the purpose of achieving, keeping or defending a condition of strategic, operational and/or tactical advantage.

(b) **The Purpose.** Causing, even indirectly, physical damage to equipment or people, or rather sabotaging or damaging in a direct way the information systems of a sensitive target of the attacked subject.

(c) **The Mean/Tool.** An attack performed through the use of technological information systems, including the Internet.

Based on the above, Stefano Mele defines a cyber-weapon as: "A part of equipment, a device or any set of computer instructions used in a conflict among actors, both national and non-national, with the purpose of causing, even indirectly, a physical damage to equipment or people, or rather of sabotaging or damaging in a direct way the information systems of a sensitive target of the attacked subject."[4]

Thomas Rid defines a Cyber Weapon as a subset of weapons more generally as computer code that is used, or designed to be used, with the aim of threatening or causing physical, functional, or mental harm to structures, systems, or living things.[5] The spectrum of cyber weapon system therefore becomes very large and hence there is a need to spell out the important facets of such weapons such as their lethality, its effect and their employment.

## Aspect of Lethality

One aspect which clearly emerges from both the above definitions is that cyber weapons are built with lethal intent although the lethality of such weapons is not quantifiable. For instance, for a nuclear strike with 20 KT weapon, the extent of damage that will be caused on any particular type of target can easily be tabulated. For a cyber weapon, such predictions can never be accurately made. How lethal is a cyber weapon, depends on its ability to cause the following to a network :–

(a) Disruption of the network services.

(b) Denial of services of the network to its users.

(c) Degradation of the performance of the network, thereby degrading the efficiency/ performance of the overall system.

(d) Destruction of critical components of the network thus rendering the entire system useless.

The easiest way to check the above capabilities of a cyber weapon would be by carrying out a live test fire, like an underground nuclear test for a nuclear weapon. However, no such tests of a cyber weapons are possible on actual networks as not only the damage caused may result in loss of property or revenue but also may be seen as an act of aggression. Stuxnet, a cyber worm attack in 2009 from an unknown origin (probably collaboration of the US and Israel) on a highly protected nuclear site at Natanz, in Iran is perhaps the only example which clearly indicated the extent of damage that a cyber weapon can cause. Although a large number of scholars till date still debate on Stuxnet as a cyber weapon, it did attack the nuclear centrifuges that operated with a Supervisory Control and Data Acquisition (SCADA) system using Siemens software and is believed to have caused a setback of at least one year if not more to the Iranian Nuclear Development Program.[6]

To establish the lethality and destructive power of cyber weapons and the network security aspects against such attacks, the US Defence Advanced Research Projects Agency (DARPA) is developing the National Cyber Range (NCR) to provide realistic, quantifiable assessments of the Nation's cyber research and development technologies.[7] The NCR will provide fully automated range management and test management suites to test and validate leap-ahead cyber research technologies and systems. It will test technologies such as host security systems, and local and wide area network (LAN and WAN) security tools and suites by integrating, replicating or simulating the technologies. Seven large-scale cyber experiments for multiple US Department of Defence (DoD) organisations were executed on the range during a one-year beta operation phase that ended in Nov 2012.[8]

## Effect of Cyber Weapons

Cyber weapons and cyber attacks offer a means for potential adversaries to overcome overwhelming advantage of a nation in conventional military power. The other most important aspect of a cyber weapon is that they need not necessarily target a military objective. In fact more often than not, the objective of a cyber attack will be a non-military target. Critical national infrastructure such as the power grid systems, telecom networks, the air traffic control system, economy sector to include banking and other financial institutions and railways could be such possible targets as all these sectors heavily rely on automation and networking. The payload of a cyber weapon could vary from a programme that copies information off a computer and sends it to an external source; or an altering and manipulating programme to either take control of the system or to alter the way in which it works. It could also be a code which could convert a computer into a botnet# and employ the machine in a Distributed Denial of Service## (DDoS) at a later stage or even cause destruction of a physical process which the computers of the SCADA system controls like it happened with Stuxnet. Various effects of a cyber weapon attacking a network have been summarised by Dr Roland Heickerö in an Effect matrix as shown at Figure 1.[9]

|  | Physical arena (land, air, maritime, space) | Information Arena | Cognative domain (cognition, perception, emotion) |
|---|---|---|---|
| Physical Effects | Interruption, destroy electronics and sensors, affect transmission and access links, derive robots, system failure | Interrupted communication, denial of services; DOS | Fragmented communication, decreased amount of information, reduced analysis capability |
| Syntax Effects | Hacking, cracking virus, Mistrust against system Trojans, spam, interception, exploit, bugging illegal misuse of information system | Attack logic of system, delay and distortion of information | Mistrust against system |
| Semantic Effects | Mass medial manoeuvers, planted information, mutilation of sensor data | Deception and manipulation of information (disinformation) | Changed situation awareness, mistrust against and questioned of information, instability for decision making |

**Figure 1: Effect Matrix of Cyber Weapons**

# A botnet (also known as zombie army) is a number of internet computers that, although their owners are unaware of it, have been set up to forward transmissions (including spam or viruses) to other computers on the internet.

## A DDoS attack os one in which a multitude of compromised systems attack a single target like a server, thereby causing denial of service for users of the targeted system.

A single cyber weapon may not be able to impact the physical arena, information arena and the cognitive domain and hence a multi-pronged cyber attack may be necessary if all the three spheres are to be addressed simultaneously. Key for success in cyber space is to understand prerequisites for conducting operations in the information arena and cognitive domain of the enemy. Cyber weapons do not create the same spectacular visual that a nuclear or even conventional missile does, which makes them weapons of stealth. Also before being hit, a network is unlikely to get any advance warning of an incoming cyber attack. Not knowing what the next attack is going to be or when it will happen has a profound effect on the victim and makes cyber weapons unique amongst all possible coercive systems. Another characteristic of cyber weapons is that for the first time in history, this technology gives small states with minimal defence budgets the capability to inflict serious harm on a vastly stronger foe at extreme ranges.

## Employment of Cyber Weapons in a Conventional War

The US Air Force has designated six cyber tools as weapons to help normalise military cyber operations and keep up with rapidly changing threats in the newest theatre of war.[10] This clearly indicates that time has now come that the armed forces around the world gear up to employ as well as defend against cyber weapons. The 2008 cyber-attacks on Georgia is the first case in history of warfare when cyber space domain attacks were synchronised by Russia with major combat actions in the other war fighting domains. Although denied till date by the Russian Government, the Russian naval and land operations against Georgia were preceded by large scale DoS attacks on Georgian Military and Government networks. The attacks continued as the Russian tanks and troops were crossing the border and bombers were flying sorties.[11] The impact: Georgian citizens could not access websites for information and instructions while the nation was being invaded.

When employed in conjunction with conventional operations, cyber weapons increase the cost of conflict for adversaries as he has to now protect the National Critical Infrastructure against cyber weapons besides the conventional and nuclear threat of the enemy. Imagine, if India was involved in a conflict situation and the adversary through pre-emptive cyber attacks crippled the movement of trains and air traffic and the civil telecom network at the mobilisation stage itself. The problems that would arise for movement of troops and logistic supplies would be phenomenal. Similarly, if the major oil refineries were to be shut down because of failure of its SCADA system, it could create a critical situation for the defence forces embroiled in a conventional war. Hence, like the lines of communication have to be protected during the war, in a similar manner in the cyber age, the Critical National Infrastructure will have to be protected against enemy cyber attacks both before the outbreak of hostilities as well as during the hostilities, thereby increasing the cost of war fighting. However, cyber weapons are unlikely to influence beyond a point the national security policy when core national interests are at stake.[12]

Like the civil infrastructure, cyber weapons can target the military networks too. Although our defence networks enjoy the advantage of being segregated from internet, however this does not make them immune to cyber weapons. Most of our defence equipment is imported and most hardware today is produced in and around China. Also, there exists no agency in the country to sanitise the network equipment when it is being imported against malware pre-embedded in them. Therefore, a cyber attack could yet be initiated during critical periods of the battle employing Embedded Devices and Trapdoors.[13] Any network in the battlefield is perhaps therefore as vulnerable to a cyber attack as a civil target is. However, developing and deploying potentially destructive cyber weapons against hardened military targets will require significant resources, hard-to-get and highly specific target intelligence, time to prepare, launch and execute an attack. Hence, the deterrence value of such weapon systems against military targets is negligible.

India too could employ such weapon systems during the preparatory as well as the contact stages of a battle to weaken the enemy's war waging potential by targeting their critical national infrastructure. After all, cyber weapons are a cheap way to build a global strike capability against networked states and armies and our potential adversaries are well on their way to create such networks, if already not existing. Therefore, the time has now come to take charge of the cyber space and develop offensive cyber capability as part of the overall national security policy. After all, like in a battlefield, even in the cyber space, offence is the best form of defence.

## Conclusion

We once lived in a world in which wars were fought by brave soldiers who faced each other in furious combat in a way that today we would find it hard to believe. In the last decade, the concept of war and war fighting has changed considerably. The massive introduction of the technology component in all spheres of warfare has ensured that cyber weapons and cyber attacks too become activities undertaken by governments to degrade the enemy's war waging capabilities.

# End Notes

1.  Definition of Weapons as given in Wikipedia

2.  Cyber Command for Country Soon: Antony TNN May 26, 2013, 03.09AM IST

3.  Thomas Rid, Cyber War Will Not Take Place, 2013

4.  Stefano Mele, Cyber Weapons: legal and strategic aspects Version 2.0, June 2013.

5.  Thomas Rid and Peter McBurney, Cyber Weapons published in The Rusi Journel

6.  Stuxnet Effect: Iran Still Reeling, Industrial Safety and Security Source, August 3, 2011, Link: http://www.isssource.com/stuxnet-affect-iran-still-reeling/

7.  The National Cyber Range: A National Test bed for Critical Security Research http://www.whitehouse.gov/files/documents/cyber/DARPA%20-%20 NationalCyberRange_FactSheet.pdf

8.  ibid

9.  Some Aspects on Cyber War Faring in Information Arena and Cognitive Domain http://www.dodccrp.org/events/11th_ICCRTS/html/presentations/157.pdf

10. US Air Force designates cyber weapons Apr 10, 2013. http://www.itnews.com. au/News/339234,us-air-force-designates-cyber-weapons.aspx

11. David Hollis, Cyberwar Case Study: Georgia 2008, available at smallwarsjournal. com

12. Ross M Rustici, Cyber Weapons: Levelling the International Playing Field http://webcache.googleusercontent.com/search?q=cache:http://strategic studiesinstitute.army.mil/pubs/parameters/Articles/2011autumn/Rustici.pdf

13. Trapdoor: also referred to as backdoors, are bits of code embedded in programmes by the programmers to quickly gain access at a later time, often during testing or debugging phase.

†   **Colonel Sanjeev Relia** was commissioned into the Corps of Signals on 20 Dec 1986. Presently, he is a Senior Research Fellow at the Centre for Strategic Studies and Simulation, United Service Institution of India, New Delhi.

# Cyber Warfare & National Security Strategy[*][†]

## Shri PV Kumar

## Introduction

Lieutenant General PK Singh, Director USI, distinguished members of the USI, members of the academia, diplomatic and press corps, ladies and gentlemen; I consider myself privileged to be here amongst you to deliver the Eleventh Major General Samir Sinha Memorial Lecture on the subject of 'Cyber Warfare and National Security Strategy'. I am thankful to USI for giving me the opportunity to speak on an issue of such contemporary relevance and importance. As Chairman NTRO, I feel a particular sense of criticality and urgency towards the issue. The topic is so vast and complex that an address of this nature may not be able to do full justice to it.

I would like to emphasise that what I say today are my personal views which do not necessarily represent the views and official position of the Government of India. While addressing today's gathering of soldiers and strategists, I will try and focus on those aspects of Cyber Warfare and National Security Strategy which would interest you. First and foremost, one is convinced that Cyber Space has blurred all conceptual and physical boundaries as we understand them in the field of warfare till now. I will briefly touch upon the evolution of society and warfare.

Human society has in the last 500 years witnessed by and large three phases of socio-technical revolution covering the periods of the 1st

* Text of the talk delivered at USI on 10 Apr 2013 with Lieutenant General Prakash Menon, PVSM, AVSM, VSM (Retd), former Commandant National Defence College, New Delhi and presently, Adviser National Security Council Secretariat (NSCS), in the Chair.

† This article was first published in the *Journal of the United Service Institution of India*, Vol. CXLIII, No. 592, April-June 2013.

and 2nd Industrial Revolutions and finally the current Information and Communication Technology (ICT) Revolution. Social structures, governance and warfare have also undergone evolution in sync with these phases of our society. It may not be incorrect to say that nothing like ICT has ever been witnessed by human society – in terms of technology itself, its impact on all aspects of our lives and the pace of change that it has precipitated.

At the core of the ICT Revolution is the ever crucial, abstract, intangible commodity – information; whether it is data, intelligence, knowledge or wisdom. Information is a virtual commodity, having some rather unique attributes. It can be shared without its value being reduced. It can be stolen and is not measurable. The same information can exist in more than one place at the same time. It is non linear in its impact; small quantities can have large effects.

## Impact of Information and Communication Technology

In today's world ICT is omnipresent and it pervades every aspect of our lives. State-of-the-art technology, ever-improving performance and tumbling costs have resulted in widespread proliferation of ICT. The ICT revolution has changed the world to a border-less entity compelling the creation of a new world order. The Internet as a network of networks has reshaped large parts of the world as a borderless world of convergence between communication and computers resulting in an unprecedented integration of peoples, structures and processes.

ICT has enabled the efficient collection and use of information. Various elements like processing machines, storage devices and communication networks etc form its core components. The large presence of ICT in public infrastructure, both critical and non-critical, has emerged both as a benefit and a threat. The technology trajectory has resulted in an ever increasing social dependence on ICT structures and mechanisms as never before. Today technology has brought ICT within the reach of the un-initiated which has resulted in its wide social impact. The Social Networking and Mass Communication systems have networked societies and individuals as never before. On the business side, ICT is today the lead money spinning industry. With these attributes, ICT and its related infrastructure have now become critical assets. And like any other national asset, ICT has become an important target for adversaries as well.

As warfare experts would like to say, ICT now forms the 5th or 6th most critical dimension of modern warfare, depending on how one differentiates between electronics and Cyber Space technologies.

## Cyber Space and Cyber Warfare

That brings us to Cyber Space, which though an offshoot of ICT, has now assumed an identity of its own. Like ICT, Cyber Space too has assumed huge proportions. ICT and Cyber Space have offered new frameworks for functional interoperability between all forms of human and Electromagnetic interactions. It is no longer a high tech venture which used to be the exclusive preserve of scientists and technologists. An important evolutionary aspect of Cyber Space is that it has emerged as a largely de-regulated medium. This has presented a new challenge to certain societal aspects which includes governance, the threat landscape and security.

Though nation states do often attempt to create regulatory structures, the Internet seems to have defeated most such endeavours, atleast till now. Like ICT, Cyber Space has become an important national asset spanning across all sectors, including governance and security. New capabilities have enabled new threat vectors for Cyber enabled warfare and Cyber Warfare itself. It has made the commission of crime easier and crime detection even more complex. While it is not my intention to over state its importance, Cyber Warfare is likely to decisively influence the pre-conflict stages and eventually the outcome of conflict itself. We need to appreciate that due to the omnipresent nature of ICT and Cyber Space, it is possible that this may lead to a new form of National and Total War.

Some distinctive features of Cyber Warfare are :-

(a) Cyber Warfare can enable actors to achieve their political and strategic goals without the need for armed conflict.

(b) Cyber Space gives disproportionate power to small and otherwise relatively insignificant actors.

(c) Operating behind false Internet Protocol addresses (IPs), foreign servers and aliases, attackers can act with almost complete anonymity and relative impunity, at least in the short term.

(d) In Cyber Space the boundaries are blurred between the military and the civilian, and between the physical and the virtual; power can be exerted by states or non-state actors, or by proxy.

Cyber Space can be viewed as the 'fifth battle space', alongside the more traditional arenas of land, air, sea and space. Cyber Warfare is best understood as a new but not entirely separate component of this multifaceted conflict environment. Warlike actions in Cyber Space are more likely to occur in conjunction with other forms of coercion and confrontation. However, the ways and means of Cyber Warfare remain undeniably distinct from these other modes of conflict.

It is said that 'war' and 'warfare' have an 'unchanging nature', yet they have a 'highly variable character': 'We know with a sad certainty that war has a healthy future. What we do not know with confidence are the forms that warfare would take. Although the concept of revolution in military affairs (RMA) is typically associated with technological advancements, it also involves changes in strategy, operations and tactics. With the dominance of information in all spheres, new strategies would keep evolving in both defence and offence.

The growing relevance of Cyber Warfare in RMA is on expected lines. At the turn of the century, the Pentagon adopted the doctrine of Network-Centric Warfare (NCW) and set out its vision of autonomous 'swarming' and 'self-synchronised' war fighting units connected to one another by high-speed data links and superior battlefield awareness. This brings us to the 'chaoplexic' form of warfare fought by decentralised networks.

Military theorists allude to the 'swarm', the networks of distributed intelligence that enable bees, ants and termites to evolve complex forms of collective behaviour on the basis of the simple rules of interaction of their individual members. Of particular interest are the resilience and flexibility of these swarms as amorphous ensembles whose continued existence and successful operation is not critically dependent on any single individual. Military swarms promise not only more adaptable and survivable forces but also new offensive and defensive tactics better suited to the contemporary battle space. Beyond the flexibility and evolutionary capability, it is also claimed that military swarms can converge on their target from all directions in offensive bursts, thereby maximising the shock effect.

Hostile actors in Cyber Space can make use of a wide range of techniques.

Malicious software (malware), networks of 'botnets' and logic bombs can all be employed to navigate target systems, retrieve sensitive data or overrule command and- control systems. Although the technology and skills involved in designing, building, testing and storing these weapons may be complex and advanced, the means by which the weapon is delivered and by which the desired damaging effect is caused may be astonishingly simple. One well-known example occurred in 2008 when highly classified US Department of Defence (DoD) networks were reportedly infected by an unknown adversary that 'placed malicious code on USB thumb drives and then dispersed them (in parking lots) near sensitive national security facilities. After a curious finder inserted the drives into computers, the code spread across their networks.

Let us examine the direct military threats emanating from Cyber Space. Cyber technology has clear military applications which can be exploited in conflict situations. Whether through military equipment and weapons systems, satellite and communications networks or intelligence data, the armed forces are highly dependent on information and communications technology. While it provides immense advantages it also throws up major challenges in terms of information overload making assessments difficult. Bombs are guided by GPS satellites; drones are piloted remotely from across the world; fighter planes and warships are now huge data processing centres; even the ordinary foot-soldier is being wired up. In a digital, knowledge-based society this is to be expected. But while technology brings opportunities it can also create vulnerabilities. The major powers have long recognised the strategic and tactical value of Cyber Space. Similarly, weaker states are now seeking to partially offset this asymmetry by developing their cyber capabilities. Military strategists have come to view information dominance as the precursor for overall success in a conflict.

## Impact on National Security and Warfare

The nature of Post-Modern Conflict has undergone a huge change, especially since the Gulf Wars and 9/11. Both state and non-state actors have achieved threat parity and terrorism is likely to dominate the conflict scenario. Cyber Warfare has emerged as an important new element of warfare. Cyber Warfare is arguably at the most serious end of the spectrum of security challenges posed by – and within – Cyber Space. Just like the tools of conventional warfare, cyber weapons can be used to attack the machinery of a state, financial institutions, national energy, transport infrastructure and even public morale.

ICT and Cyber Space have had a profound effect on the affairs of the state. Free flow of information across TV screens, e-mails, cyber chat rooms etc contribute to wider event awareness, debate and transparency. Who could have imagined the Arab Spring and Shahbagh movements 20 years ago. But the Information revolution generates its own contradictions. It strengthens forces of both anarchy and control.

From what we see today as part of youth movements powered by Social Media, the individual has become more empowered as compared to social and government structures. Many hierarchies lie destroyed and are being replaced by new and more broad based power structures. The ICT revolution also offers too many choices, greater insight and has the potential to increase the fog – both in peace and wartime. As was brought out earlier, ICT or Cyber Space structures have become vital national assets.

The National Information Infrastructure, including computers, networks, storage devices, communication systems, cyber enabled and cyber controlled systems etc, has assumed an importance unheard of before. As has been shown in many Hollywood movies, cyber linked physical infrastructure is now a genuine target of physical destruction or disablement through cyber means. Hardware is just as susceptible as software. Backdoors and malicious code or circuitry hidden inside counterfeit hardware and software, all the way down to the basic input-output system (BIOS) and instructions set inside the integrated circuit chips is a case in point. Any vulnerability in the BIOS of microprocessors can be exploited to gain control over the computer. The design, manufacturing and testing stages of IC production are done in a diverse set of countries. This makes quality control a difficult proposition. With commercial off-the-shelf (COTS) procurement and global production, there is an increasing risk of covert hardware/firmware based cyber attacks. Most of us know what a digital worm or a virus like Stuxnet can do to the physical world of Supervisory Control and Data Acquisition (SCADA) controlled systems. Aviation, railways, power systems, food supply chains, R&D facilities, e-governance structures are today vulnerable in a threat mosaic never encountered earlier.

ICT's impact on financial systems including banks, stock-exchanges, electronic fund transfer mechanisms, e-commerce architectures etc has resulted in new threat and security frameworks. Defence assets, structures and related vulnerable areas and vulnerable points are under severe threat today, both during peace and wartime.

Let us now look at what is under threat in the world of ICT itself. The Internet population has jumped from 1.15 billion users in 2007 to 2.27 billion in 2012; i.e. it has almost doubled in five years. The largest and fastest rising numbers are in Asia with India and China at the top is no surprise. In an Internet minute, nearly 1 terrabyte of data is shipped, 1300 new mobile users added, 204 million e-mails sent, more than 6 million Facebook views generated, more than 2 million search queries on Google, 62000 hours of music transacted, 30 hours of video uploaded and 1.3 million video views generated on Youtube.

For people of my generation this sounds mind-boggling. But let us get shocked further. Today the total number of networked ICT devices equals the world population. By 2015 these will be twice the world population at that time. Today Global IT revenues have exponentially jumped from USD 350 million in 1997 to around 120 billion in 2012. In all, including services, telecom etc, ICT can be valued at around 6.8 trillion dollars.

## Issues

The Internet technologies, which employ open standards for exchange of information and have made this mind boggling scale of things possible, are not fundamentally secure. This fact, as a result of which systems remain ab-initio vulnerable, needs to be appreciated when studying the security aspects. The systems were made even more vulnerable due to compromises affected for commercial convenience and making them user friendly.

The threat to information infrastructure today spans processing elements, storage systems, transmission networks etc. On the other hand, easy availability of information to the adversary poses a challenge which has to be dealt with without affecting own systems. Today cyber enabled Information Warfare has further enhanced the threat to the decision making process through more efficient information disruption and misinformation mechanisms.

## Social Media

The grievances in the Gaza War may be ancient, but some of the weapons reportedly being used are spanking new; reflecting the changed nature of war in this cyber era. One reads for example about a part of the war being tipped in a side's favour based on the number of Twitter posts far outnumbering those of the other side. Recently in Egypt and Libya, massive riots were led by extremists who were apparently united and who stormed embassies. Riots and demonstrations followed all over the world. It has been reported that the attacks on the embassies were executed in a coordinated manner on multiple embassies at the same time. The attacks were reportedly incited, spread and well coordinated through social media like YouTube, Facebook and Twitter.

Social media seemed to have been employed to stoke an insurgency. It illustrates how, often something innocuous can be get blown out of proportion by certain powers with an agenda using this new weapon in their arsenal. This level of social manipulation can be readily adopted by foreign powers to foment trouble well outside their own national borders. The magnitude, scale, apparent-spontaneity, decentralised nature yet well networked and coordinated nature of this attack - seem to fit well with the theories of 'Chaoplexic Warfare' mentioned earlier.

This may be the right moment to take a peep at the exotic world of Cyber threats and terms like hacking, phishing, Denial of Service Attacks, Botnets etc. Without going into jargon, these are various forms of threats and delivery systems. For example, as is apparent from the term, Denial of Service (DoS), is actually denying users a service mostly through inundating the Service Provider.

Botnets are groups of zombie computers under the control of a remote and invisible hacker forced to function in a manner not desired of them. Today hackers can control an army of bots all over the world which can be used to attack a system, a network or networks, a service or a nation. Can these zombie army of bots, involving your and my computers, be used to attack a nation. Yes, it is possible. Remember Estonia, where an entire nation was paralysed by a cyber attack. Estonia happened to be one of the most wired countries at that time.

When these bots are used to fire large cyber traffic to inundate an entire target infrastructure the DoS becomes Distributed DoS. This seems to have happened to the US banks recently. You may recollect what happened last month to the Spamhauss organisation which deals with anti-Spam operations. It was probably the largest cyber attack which has come to notice with thousands of Bots initiating millions of transactions on Spamhaus servers, effectively shutting them down and even slowing down the Internet. We all know how Stuxnet was used to disable the Iranian nuclear programme attacking the Siemens control systems deployed for operating centrifuges. Or for that matter the subsequent Flame virus which was focused on Middle East for information collection. Around 30000 bots were used to target Aramco of Saudi Arabia which is one of the largest oil producers. The production there was disrupted for many days.

## Attribution and the China Bogey

The People's Republic of China (PRC), in particular, has long recognised the strategic and tactical value of Cyber Space. The PRC is believed to be focusing inter-alia on its cyber capabilities to counter the current military asymmetry with the USA in terms of military and technical hardware. Chinese military strategists have come to view information dominance as the precursor for overall success in a conflict.

A lot is written about China's prowess in the field – this is a possibility that cannot be denied given the fact that China has emerged as a formidable force in the world of technology. The Titan Rain attacks in 2007 – one of the most large-scale infiltration of the US and UK government departments, including the US DoD and the UK Foreign and Commonwealth Office – were attributed to China, and had allegedly been under way since 2002. Furthermore, in Mar 2009 China was linked to 'GhostNet' when it was revealed that a large-scale spying network had attacked a significant number of government departments and strategic targets, including the Tibetan community.

On 19 Feb this year, a report by the US-based Cyber Security firm Mandiant accused the Chinese military of being behind a series of cyber attacks against businesses, institutions and infrastructure in the US. That was not the first time that China received accusations of this type, although the novelty was that the report localised in detail the origins of the attacks. According to Mandiant, a Chinese army building in a suburb of Shanghai was responsible for most, if not all of the attacks.

Beijing categorically denied the charge adding that it is also the victim of numerous attacks, which have increased over the years and most of them are from the North American country. A computer security official said China had become the world's biggest victim of cyber attacks, with a report from a national computer monitoring centre revealing that many domestic computers were controlled via overseas-based IPs last year. A total of 47,000 overseas IPs were involved in attacks against 8.9 million Chinese computers last year, compared to nearly 5 million affected computers in 2010, according to a report issued by the National Computer Network Emergency Response Coordination Centre of China (CNCERT/CC), China's primary computer security monitoring network. Most of the IPs originated in Japan, the US and the Republic of Korea (ROK), according to the agency. Since attribution is very difficult in the cyber domain, it is difficult to conclusively support the argument that China is actually behind much of what is being witnessed.

## Indian Cyber Scenario

In India, as we are all aware, there is a near total reliance on external sources for hardware and software (operating systems, application software, antivirus, network protocols et al). In view of this, it is virtually impossible to have complete information on hidden vulnerabilities such as the presence of trap doors and malware.

Some mitigation strategies could include those most essential for resilience, namely a full understanding and control over the technologies and systems of the critical infrastructure, cyber security awareness and education, sanitisation techniques and strong cryptography, good security-enabled commercial information technology, an enabling global security management infrastructure and a strategic push to investment in indigenous development/ production of hardware and software. This needs a focus on Research and Development particularly in areas of : authentication technologies; secure fundamental protocols; secure software engineering and software assurance; holistic system security; monitoring and detection; mitigation and recovery methodologies; cyber forensics: catching criminals and deterring criminal activities; modeling and testbeds for new technologies; metrics, benchmarks and best practices.

## Technology and R&D

Cyber technologies are a very new field as compared to other established technology areas. These technologies have matured over the last two decades or so. Hence there is a large need to develop not only secure cyber frameworks but also put in place defensive cyber technologies to guard against various threats which were mentioned earlier. It is an understatement to say that a fast moving R&D framework is the bed rock of any cyber security endeavour. As has been enumerated in some of the international cyber security documents, we require to develop R&D strategies to enhance the security, reliability, resilience, and trustworthiness of digital infrastructure. The R&D effort also needs to be coupled with technology forecasts to cater for immediate and long lead items.

There is a pressing need for developing technologies for various segments which include secure data communication technology, encryption frameworks, resilient and sustainable digital infrastructure, large scale realtime data processing systems, threat identification and intrusion detection systems etc. While the technologies, I just mentioned, have more to do with enabling mechanisms, the cyber defence and mitigation strategies require a parallel development effort to cater for containing the fallout of major cyber attacks. In addition, this also needs constant evolution to keep pace with the new types of cyber threats which are encountered on nearly hourly basis.

While development of technology has a dynamics of its own, a realistic appraisal of the available technologies and their likely trajectory, needs to be undertaken in a comprehensive manner. Such forecasting is required both for the enabling systems and from the cyber warfare point of view. This also includes constant and independent analysis and assessment of vulnerabilities in the existing systems, hardware, networks, processing elements etc. This is a very challenging task in its own right. A word about who needs to do all this.

Based on what we understand of the all pervasive nature of cyber space and its effect on all aspects and dimension of our societies, it is not possible for any agency, sector or government to go it alone. It is inescapable that various arms of government, academia and industry collaborate and coordinate efforts to cater for the demands of such a fast changing field. It can be appreciated that entities both within the government and outside have strengths of their own which are best suited to take on the respective areas of responsibilities.

## Cyber Security Manpower

I would also like to emphasise on a very important related aspect. While we have talked so much about cyber technologies, its potential and dangers, it needs to be appreciated that it is finally the man behind the machine which is the most important factor for any success, especially in the cyber world.

The developments during the last one to two decades indicate that while sufficient expertise is available in the field of exploiting cyber technologies, it is the cyber security expertise which is not available in the required numbers as yet. In addition, such expertise is also not available with the required capability levels. We need a constant supply of capable man power from academic institutions and industry to pitch in for defending Cyber Space. It is a long haul but an institutionalised start needs to be made; otherwise while developers would have created systems with enormous capabilities, defending such systems will lag behind which could result in a potentially disastrous situation.

## Secure Technology Framework

The panacea for such a state is indigenisation and this is being aggressively pursued. We need to be able to develop world class products (operating systems, application software, hardware such as network components, even chips being used etc) that we can use with the kind of faith that comes from knowing everything about it. Greater awareness of Cyber Security aspects through training, information dissemination, adoption of best practices, regular cyber audits by experts etc would also contribute significantly. This aspect is of great strategic importance and needs active involvement of academic institutions, industry, think tanks and government institutions.

Cyber Space is also a global medium and we need to partner with our friends across the world. While some measures have already been initiated, India needs to actively participate in the international Cyber Security dialogue to safeguard our interests.

## The Challenge

The scale of ICT applications and their openness which is conducive to growth, throw a sort of 'grand challenge problem' in protecting cyber assets from penetration and attacks. Cyber-attacks are now becoming the stuff that we read of at breakfast in our morning newspapers. In this regard I have already given examples of incidents reported in the recent past. A significant area of concern is cyber espionage which is the most prevalent of the cyber activities. Whether used to uncover sensitive government information, steal trade secrets or commercial data or as part of intelligence or reconnaissance work, it fits into the doctrine of using 'information superiority to achieve greater victories at a smaller cost'.

Many nations are pursuing offensive cyber capabilities, but not much is revealed about this in the public domain. However, in a recent departure from this norm, Chief of the US National Security Agency recently disclosed that the US DoD is establishing a series of cyber teams to combat the threat of a digital assault that could cause major damage and disruption to the country's vital infrastructure. He mentioned that 40 teams should be ready by 2015 and that 13 of the teams will be offensive fighters specifically designed to attack other countries while the other 27 cyber teams were being established to support the military's warfighting commands. Some others will protect the Defence Department's computer systems and data.

There can be a psychological dimension to cyber-attacks. The infiltration, of what are assumed to be secure systems and critical infrastructure, highlights national vulnerabilities and weaknesses. This can provoke feelings of insecurity which could indeed be the attacker's goal; in the same way that the fear of terrorism and its potential harm can have a detrimental and disabling effect almost as great as the terrorist act itself.

Developed countries frequently complain about large scale espionage and surveillance operations by cyber attackers, with their defence and hi-tech industry as one of the prime targets. In the case of suspected state-sponsored actions it is difficult to establish beyond any doubt that the order to attack originated in the executive or presidential office, let alone a capital city. Furthermore, the difficulties of attribution allow a degree of plausible deniability. Perpetrators can cover their own tracks and implicate others, particularly when third-party servers and botnets in unrelated countries can be used to originate attacks and provide cover for the actual attacker. The increasing integration of National Information Infrastructure with military information infrastructure has diffused the boundaries between civil and military information networks.

Can we imagine what will happen to us in a situation like Estonia, the US or Saudi Arabia? Do we fully understand the threat? Do we have a measure of it, and are we prepared for it? Are we still working with archaic civil-military frameworks? Is there a difference between peace and wartime? Are we matching up with the faster information proliferation and propagation mechanisms leading to information flow management problems? Are our business models geared for the threat? Does the industry and the government's L-1 system of procurement provide for the commercial vs security compromise? Do security overheads not often tempt us to opt for relatively unsecured but operationally capable systems for ease of operation and management? All I can say is that policy frameworks are in the process of being put in place to address such issues.

## Policy Options

Never in the history of national security management have such high demands been placed on information collection mechanisms which need to process such large amounts of data and at such high speeds. A 24x7 situational awareness and matching fast paced mitigation mechanism can not be delayed any longer. However, this may agitate some privacy advocates who may be justified in treating national security threats and privacy at par. I would like to bring out that there are enough comprehensive frameworks available in Indian laws and the IT Act to arrest such negatives. Similar examples are available across the world like the legal formulations in the US and the UK. The European Union is also evolving its own version for a multilateral and multi-nation framework. It may require a full discussion to dwell upon the various dimensions of the legalities in ICT.

Recently, the US had initiated a series of policy measures with their President pushing it past the Congress. There could be lessons for us also. The policy framework needs to address the immediate and futuristic requirements; as the threat is here and now our adversaries, whether state or non-state actors, are already on the job. A national level coordination effort at policy and operational levels is the foundation layer for any cyber security endeavour. India needs a national cyber coordination mechanism for threat assessment. This should have multi-agency participation. Some efforts in this direction have already been initiated. This endeavour will result in a credible Information scanning framework in coordination with the service providers and industry.

In order to make sense of it all, a comprehensive data analytics capability with mining and fusion mechanisms needs to be put in place for predictive trend analysis. Such skill sets need to be honed with advanced simulation and modelling techniques. A large body of research work already exists in this area which is waiting to be absorbed. While information collection and predictive efforts fall in the pre-event category, measures will require to be put in place for the post-event phase also.

Its time we touch upon the mitigation strategies also. This will involve counter measures and realtime forensics. The mitigation and counter measure aspects need to be handled in a coordinated manner at the national level. We need to prepare a Cyber Security Incident Response Plan and enhance public-private partnerships. While so much needs to be done at the operational level, there exists a very large need for cyber threat awareness, both within and outside the government. There is a need to initiate a national awareness and education campaign to promote Cyber Security. This basic measure needs to be undertaken at multiple levels of society and governance.

## National Centre for Critical Information Infrastructure (NCIIPC)

Now I will briefly mention about the NCIIPC. The amendments made to the Information Technology Act in 2008 reflected the nation's recognition of the need to adopt an institutional approach to enhancing our cyber security profile as also to take steady but sure steps towards protection of its critical information infrastructure. The IT Act envisaged the creation of a specialised body to synergise our collective efforts and capabilities to protect the Nation's critical information infrastructure.

Critical sectors as you are all aware, are those sectors whose incapacitation or destruction would have a debilitating impact on national security, economy, public health or safety. Several initiatives have been launched in recent times to enhance our cyber security profile. The creation and activation of the NCIIPC for protection of our critical information infrastructure is one of the important components of this construct.

## Conclusion

In conclusion, I may say that Cyber Space offers mind boggling opportunities for improving the quality of life and work but it also provides a threatening landscape for destroying it. There is no escape from Cyber Space or its threats. Society, visionaries, technologists and the Cyber Space 'Subscribee' citizen need to pitch in. Many old paradigms are no longer relevant. New frameworks need to be put in place with no loss of time. The Cyber world is all about speed. We can't afford to be left behind.

---

‡ **Shri PV Kumar** is a career intelligence officer. He joined the Research and Analysis Wing (R&AW) in 1971 and held important appointments in that organisation, both within India and abroad. He joined the National Technical Research Organisation (NTRO) in Oct 2009 as Senior Adviser and Deputy Chief of the Organisation. He took over as Chairman, NTRO on 31 Mar 2011.

# Cyber Weapons – The New Weapons of Mass Destruction?*

## Lieutenant General Davinder Kumar, PVSM, VSM and Bar (Retd)

### Introduction

In this Information Age, while Cyber Space has been accepted as the new domain of warfare, it still does not have an internationally accepted definition. The same is true for cyber weapons. The US Government security expert, Richard A Clarke, in his book, Cyber War defines Cyber Warfare as actions by a nation state to penetrate another nation's computers or networks for the purpose of causing damage or disruption.[1] Cyber Weapons, accordingly, are the tools for conducting cyber warfare.

Weapons, in their simplest form, could be considered as, "instruments of harm". Since the dawn of time, humans have used weapons to hunt and demonstrate or acquire power. The types of weapons and their range, lethality and precision have increased substantially with the advancement of technology and the need to obviate the perceived threat. The weapons thus have evolved with time. The time taken to translate the concept into a product/weapon has been reducing in consonance with the pace of development of technology and its engineering into production. Cyber weapons are also evolving just as the conventional weapons albeit at a much faster pace. In the cyber world, the technological advancements happen in days or even hours with the emergence of corresponding new threats. The most significant development has been the reach of cyber weapons in the real physical world as demonstrated by the Stuxnet attack on the Iranian Nuclear facility.[2]

---

* This article was first published in the *Journal of the United Service Institution of India*, Vol. CXLII, No. 591, January-March 2013.

## Definition

We need to define cyber weapons correctly as the same has significant political, security and legal consequences. This is an urgent and important requirement - for being able to assess both the level of threat from a cyber attack and the consequent political and legal responsibilities attributable to the attacker. Two definitions, one by a security expert and the other with legal overtones are given below:

> *"Cyber Weapon could be defined as a computer code that is used or designed to be used, with the aim of threatening or causing physical, functional or mental harm to structures, systems or living beings."*

> *"A device or any set of computer instructions intended to unlawfully damage a system acting as a critical infrastructure, its information, the data or programmes therein contained or thereto relevant, or even intended to facilitate the interruption, total or partial, or alteration of its operations."*

The above definitions imply that cyber weapons may span, in theory, a wide range of possibilities: from Denial of Service attacks (which typically have a low level of penetration) to, "tailored" malware like the Stuxnet characterised by high intrusiveness and a low rate of collateral damages. It may be prudent, therefore, to evaluate cyber weapons from its domain of relevance, cyber space, with the distinct possibility to cross the virtual boundaries and extend to the real world.

## Classification[3]

With the above idea in mind, cyber weapons can be classified according to the following four parameters as shown at Figure 1 :–

(a) **Precision** - That is the capability to target only the specific objective and reduce collateral damages.

(b) **Intrusion** - the level of penetration inside the target.

(c) **Visibility** - the capability to remain undetected

(d) **Ease of Implementation** - a measure of the resources needed to develop the specific cyber weapon

## Classification Of Cyber Weapons

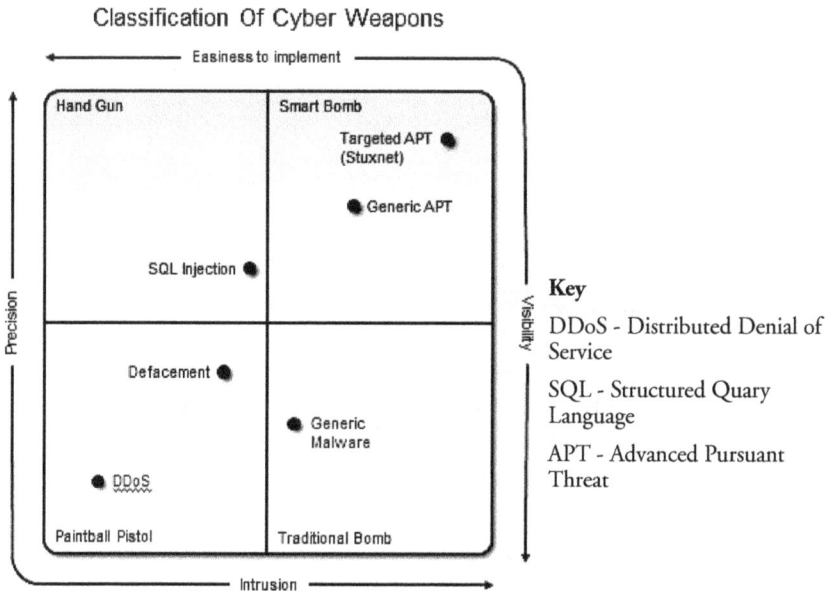

**Key**

DDoS - Distributed Denial of Service

SQL - Structured Quary Language

APT - Advanced Pursuant Threat

## Why Cyber Weapons? [4]

Use of cyber weapons is complementary to conventional military strikes. It could be possible to:-

(a) Support offensive operations by destroying enemy's defence/critical infrastructure

(b) Probe the technological capabilities of the adversary by evaluating the ability of an agent to infect the enemy system.

(c) Cyber weapons are more efficient and less expensive.

(d) The attack is carried out at the speed of light.

(e) Cyber weapons are less noisy (stealth weapons) - no one wants to acknowledge the vulnerabilities of their system.

(f) Attribution is very difficult - the possibility to operate under cover makes cyber weapons very attractive.

(g) Cyber weapons are offence dominant and ideal weapons for asymmetric warfare - the warfare of 21st century

(h) Preparation phase of cyber weapons is easy to hide from prying eyes and development of cyber weapon is hard to identify.

The above advantages make cyber weapons very attractive to those "small" states that despite having reduced funds for military expenses can compete with the most powerful countries in the new domain. At present, nearly 140 countries in the world are engaged in the development of offensive cyber warfare capability.

## Likely Targets of Cyber Weapons - Impact on Cyberspace [5]

The spread of a malicious agent in cyberspace could lead to the loss of human lives and cause the destruction of critical infrastructures. These are considerable as a direct effect for which the cyber weapon has been designed, but there are also "collateral damages" caused by the uncontrolled diffusion of a cyber weapon. A cyber attack could cause similar damage to a conventional attack with serious impact on citizens. The spectrum is very wide. In general, cyber weapons could hit every critical infrastructure and vital systems of a country such as:-

**Electronic National Defence Systems.** By hacking a defence system of a country it is possible to control its conventional weapons, for example there is the possibility to launch a missile against the state itself or other nations. Similarly, Command and Control systems of the adversary can be degraded substantially by interfering/disrupting the defence communication networks

**Hospitals.** Electronic systems present in hospitals and health centres could be exposed to cyber attacks that can compromise their functioning, causing serious consequences.

**Industrial Control Systems (like Supervisory Control and Data Acquisition (SCADA) System or Programmable Logic Controllers (PLC) of Critical Facilities.** A cyber attack could compromise the management system of a chemical plant, dams, energy production plants or a nuclear site, altering production processes and exposing large areas to risk of destruction.

**Water Supply.** Water is an essential resource for the population. Interruption of the supply might leave large areas without water. The alteration of the control system might allow it to be functional but vulnerable to a successive attack such as water poisoning.

**Fully-automated Transportation Control Systems and Civil and Military Air Traffic Controls.** All those systems do not require conductors or drivers, or give a sensible aid to the conduction and control of transportation. Consider the effect of an attack on train control systems or to an air traffic management system.

**Electricity Grid Management Systems.** This target represents the vital system of a country. Attacking these systems, it is possible to interrupt the electricity supply, causing the total block of the activities of a nation such as computers, trains, hospitals and telecommunications services. These represent a privileged target for a cyber attack, and their defense is a fundamental in every cyber strategy.

## Communication and Data Networks

Banking Systems and Financial Platforms. Financial systems are critical assets for a nation and their blocking could cause serious problems. Despite being unable to cause the direct loss of human lives, a cyber attack could cause the financial collapse of a nation through interference/blocking of all economic activities. The scenario is worrying - if we think that global finance today is strictly dependent on the economy of each single state, a cyber attack against a state could cause serious and unpredictable consequences to the entire economic system.

## Limitations of Cyber Weapons

One of the most dangerous effects of the use of a cyber weapon is the difficulty to predict its diffusion since cyber space has no boundaries. This means : –

(a) Cyber weapons could hit in unpredictable ways other systems and networks that are not considered targets. In extreme cases, there is a possibility that it attacks the systems of host nation in a sort of "boomerang effect".

(b) Presence of cyber weapons in cyberspace could open up the possibility of reverse engineering of its source code by ill-intentioned individuals. Foreign governments, cyber terrorists, hactivists and cyber criminals could be able to detect, isolate and analyse the agents, designing and spreading new cyber threats that are difficult to mitigate.

(c) Cyber weapons have limited shelf life since they are designed to exploit a particular vulnerability.

(d) Span of attack and extent of damage is inversely proportional to the sophistication of the cyber weapon

## Generations of Cyber Weapons[6]

Cyber weapons, like any other weapon system are evolving with time, technological advances and threat perception. The three generations of cyber weapons could be defined as follows:

**Generation 1.** Physical or (Anti) radiation electronic warfare weapons that can blind, cripple, degrade or incapacitate through physical attack or traditional electronic warfare means. These are effectively command and control weapons. The criterion is the level of effect delivered. Traditional effects are degradation, disruption of communication with very closely controlled deployment and targeting. Examples are the blowing up of the Siberian oil pipe line in 1982; destruction of Iraq's power grid by deploying carbon fibers to short the electric grid; blowing up of Baghdad telephone system in the Gulf war and so on.

**Generation 2.** Software and hardware derived technical implementations that allow for vulnerabilities to be exploited in the systems of systems or specific targets. These are characterised by their requirement that somebody has an exploitable feature in systems design, configuration, or software implementations. This is further characterised by heavy reliance on network infrastructures though they may not be the primary mechanism of exploitation. There is varying levels of barrier to entry. Traditional characteristics are of espionage and sabotage with varying level of sophistication and control of deployment. Estonia[7] and Georgia[8] incidents would qualify.

**Generation 3.** Fusions of generation 1 and 2 weapons then become point and shoot weapons that can destroy, degrade or disrupt the adversary's systems without requiring the vulnerabilities to be exploited. The adversary is no longer required to make a mistake. These kinds of weapons simply destroy the command and control, (communication and coordination) behaviours of cyber infrastructures. Emerging characteristics are of selective targeting and speed of deployment.

Generation 1 weapons primarily work against the availability of systems and the inherent infrastructures that they operate upon. Generation 2 weapons tend to operate at the logical layers against the protocols and applications that run on top of the network. Finally, Generation 3 weapons appear to be destined to work against the entirety of the systems of systems infrastructures inclusive of the human being.

Another interesting way to define cyber weapons would be based on the contemporary threats and their classification.[9] The current state of threats is best represented as a pyramid. The base of the pyramid is made up of all kinds of threats – what we call 'traditional' cybercrime. Its distinguishing features include a reliance of mass attacks targeting ordinary users. Cyber criminals' are mostly interested in launching these attacks for direct financial gain. This accounts for over 90 per cent of all contemporary threats.

The second tier is made up of threats aimed at organisations. These are targeted attacks, which include industrial espionage, as well as targeted hacker attacks designed to discredit their victims. The attackers are highly specialised and work with a specific target in mind or for a specific client. The goal is to steal information or intellectual property. Financial gain is not the attackers' primary goal. This group of threats also includes a variety of malicious programmes created by certain companies at the request of law enforcement agencies.

The third tier, which is the top level of the pyramid, includes malware which can be categorised as true cyber weapons. These include malware created and financed by government-controlled structures. Such malware is used against citizens, organisations and agencies in other countries.

To summarise, we can identify three main groups of cyber weapons based on threats:-

(a) **Destroyers.** These are programs designed to destroy databases and information as a whole. They can be implemented as 'logic bombs' that are introduced into victim systems either in advance and then triggered at a certain time, or during a targeted attack with immediate execution. The most notable example of such malware is Wiper.[10]

(b) **Espionage Programmes.** This group includes cyber weapons (malware) like Flame, Gauss,[11] Duqu[12] and miniFlame[13]. The primary purpose of such malware is to collect as much information as possible, particularly very highly specialised data (e.g. from Automated Computer Aided Design (Autocad) projects, SCADA systems etc.), which can then be used to create other types of threats.

(c) **Cyber Sabotage Tools.** These are the ultimate form of cyber weaponry – threats resulting in physical damage to targets. Naturally, this category includes the Stuxnet worm. Threats of this kind are unique and require adequate intelligence and R&D resources. Some developed and networked countries are devoting more and more effort to developing this type of threat, as well as defending themselves against it.

## Cyber Weapons – The New WMDs[14]

Just as the industrial revolution brought about a fundamental change in warfare, the Information Age is ushering in a new, low cost option for strategic defence in the form of cyber warfare in general and cyber weapons in particular. These can now accomplish most of the strategic tasks that once required air superiority or nuclear capability. The situation is similar to the time when early nuclear theory wrestled with many of the similar issues that we now face in attempting to understand cyber weapons. Some of the important issues are:-

(a) The long range strike capabilities of cyber warfare have the potential to be extremely effective when employed as an anti-coercion weapon (power projection capability at minimal cost).

(b) A strong cyber capability is a deterrent force that will largely mitigate outside interference in domestic and regional affairs.

(c) Cyber weapons have the potential to become an equalising force because they require a fraction of investment compared to nuclear weapons or the strategic air power and yet would be able to execute most of similar missions and that too with limited or no collateral damage. While a cyber weapon can cause a total black out of the electric grid for the operationally desired duration, the same can be restored just by a click of a switch!

(d) Given the speed and precision with which a cyber attack can be carried out, these weapons can be used for anything from a warning shot to signal an adversary to a catastrophic strike that could cost trillions of dollars and an unspecified discomfort to the people.

The wide range of issues mentioned above make the cyber weapons unique. The fact that a cyber arsenal is also exceedingly cheap means that the available destructive capacity for poor and weak States vis-à-vis a developed and networked State is unprecedented. The ability to strike quickly and on such a scale with no possibility of retribution makes cyber weapons uniquely terrifying. A well executed cyber campaign coupled with careful public relations has the potential to traumatise a society in ways not seen after Nagasaki. Cyber weapons are a cheap way to build a global strike capability against networked states.

## Implications[15]

Curtailing of inter-state coercion. Just like large and capable conventional forces, cyber weapons present a strong deterrent for a potential attacker. While very few countries have the capability of intervention at the regional or global levels, any country with a network connection may be able to launch an effective retaliatory strike. Consequently, interventionist foreign policies will become exceedingly expensive both in the material and human cost. The new dangers that the fifth domain of warfare creates will limit the behaviour of bigger nations. There is a school of thought that the Iraq war would have to be fought differently if Iraq had cyber weapons.

**Derailment of Human Security Issues.** The cost of intervention increases in direct proportion to the target state's ability to launch a strategic cyber attack. Accordingly, not many nations would like to intervene to prevent humanitarian crises.

**Alteration of Conventional Force Structure.** Cyber weapons present the possibility of altering conventional force structure in a fundamental way. For example, there are multiple comparative advantages of cyber weapons over air strikes. The first and the most compelling is cost. The second is the temporary nature of the effect of cyber weapons and finally negligible collateral damage caused by cyber weapons.

## Cyber Deterrence

The above implications mean that cyber deterrence is capable of reducing the incidents of violence in the International system. At the same time, it is also likely to make the world a safer place for corrupt and abusive regimes. Cyber weapons and their value may not rival that of nuclear weapons at present, but they certainly have greater deterrence force than conventional systems. These have the potential of increasing the transactional cost of war to such an extent that developed nations will be far less willing to use force internationally based on ideals or a perception of marginal regional balance of power.

## Failure of Deterrence[16]

There are, however, glaring issues regarding deterrence in cyber space and these are:-

(a) Unlike nuclear weapons or any other conventional capability, it is almost impossible to demonstrate cyber power.

(b) It is very easy to develop this capability with an exceedingly small foot print.

(c) The technical nature of cyber weapons requires a pre-existing vulnerability in the software or the ability to assume the identity of a trusted user to carry out an attack (Identity theft when viewed in this context becomes extremely serious).

(d) The shelf life of a cyber weapon is limited to the presence of the particular vulnerability. Further, a near perfect defence against a cyber weapon can be brought in a matter of days or weeks against the use of that particular exploit.

(e) Cyber weapons, at present, can only penetrate network defences if there are exploitable flaws in those defences (It is in this context that probing attacks on the networks are to be viewed seriously and reported).

(f) Most of the technology and material needed to develop sophisticated cyber weapons are commercially available and completely unregulated. Consequently, traditional technology and arms control regimes are impossible to create and verify.

(g) Currently, the only way we can correctly estimate the cyber capabilities of another actor is by measuring the frequency and sophistication of attacks emanating from a source/state.

The world has dealt with the threat of Weapons of Mass Destruction, commonly known as WMD, in the past. However, in the world of Cyber

Space, we are now confronted with a new WMD threat: Weapons of Mass "Disruption".[17] If we do not prepare now, we could, one day, face a cyber attack that could cripple our government, our economy and our society. We need to formulate and announce our Doctrine regarding Cyber Warfare and develop demonstrable Cyber Defence and Cyber Attack capabilities to act as an effective deterrent.

In the nuclear domain, the nuclear weapon states have promulgated their respective doctrines and have built or are in the process of building a strategic triad of land, sea and airborne capabilities of launching nuclear weapons. Consequently, the nuclear deterrent has held thus far. In the Digital Age, we need a, 'Cyber Triad'[18] that will deter cyber attacks on our information infrastructure by employing Weapons of Mass Disruption. The various legs of the 'Cyber Triad' are as under : -

(a) **Resilience (First Leg).** Cyber resilience would mean such things as Redundancy of critical connectivity; the ability to handle increased traffic loads under the most stressed conditions; and the ability to protect and secure sensitive and private information.

(b) **Attribution (Second Leg).** Our continued inability to attribute attacks tantamount to an open invitation to those who would like to harm us, irrespective of their motives. If the adversary can attack our networks and systems without leaving finger prints, they can attack without consequences and that means they cannot be countered or deterred; a serious matter indeed. To deter cyber attacks, we need to improve our capability to attribute these attacks to their ultimate source and display a very strong political will that we will respond in the most devastating manner. Concurrently, we need to announce our likely response to probing or cyber espionage attempts by any player.

(c) **Offensive Capabilities (Third Leg).** Just as in the kinetic weapons, the adversary must know that the nation has an effectively balanced defensive and offensive cyber capabilities backed by a very strong political will. The nation's strategic doctrine must clearly state both the likely response to a cyber attack and the response that the same would invite. For example, many countries have promulgated that an electronic attack on their assets will be construed as an act of war and would attract appropriate response.

95

Development of credible cyber deterrence would have to be a national effort that would involve the government, industry, academia and most of all the people. At the same time, we will have to work very closely with the international community to develop a forum for peaceful co-existence in the cyber space, the corresponding legal framework; the development of secure products and services and the likely international response to a cyber attack on a nation state. We need to work harder on fostering international alliances and be an active member in the formulation of best practices and a code of conduct in the cyber space.

## Conclusion

Information and Communication Technology (ICT) has created a virtual world with no boundaries where the rules of engagement are being constantly defined by the world community with very little interference from the government. This virtual world called the, "cyber space" has not only opened new avenues for interaction, development and exchange of views but as a natural corollary, it is also becoming more hostile. Worldwide, people and in some cases, the governments are engaged in the exploitation of the cyber space for illegal activities like espionage, theft of technology, financial frauds and so on. They have, accordingly, developed means and methods to carry out such activities by way of viruses, root kits, malware and so on. These are the initial steps in the evolution of cyber weapons which till date do not have a formal definition. This evolving threat to society as a result of these cyber weapons and their capability to disrupt networks, systems and their functionality; their suitability for the conduct of asymmetric warfare, coupled with the all pervasive application of ICT in the Military and civilian domains, have opened a new dimension of warfare.

The recent critical development wherein ICT has enveloped the physical space has led to both the weaponisation of cyber space and the consequent threat to the Critical Information Infrastructure (CII). While blowing up of the Siberian oil pipeline in 1982 indicated the possibility; the destruction of Iraq's power grid by deploying carbon fibres to short the electric grid in Gulf war, the interference of financial system in Estonia and effective neutralisation of the war fighting capabilities in Georgia displayed the emergence; the Stuxnet attack on the Iranian Nuclear facility and the recent discussions in the Pentagon of possible option of taking out the Libyan Air Defence system by cyber weapons are pointers to both the emanating threat and coming of age of cyber weapons. While the USA, Russia, China, North Korea and Iran are said to have developed effective cyber weapons, many other nations are engaged in developing the same. The challenge is to stop their development and proliferation or at least regulate them through generating trust and confidence. There is an urgent need to have international treaties for limiting the use and proliferation of cyber weapons, independently or as part of expanded role of the United Nations. India, while participating in the international efforts, must enter into alliances - bilateral, multi-lateral or regional to secure her national interests and sovereignty. She must formulate a national doctrine and develop credible cyber offensive and defensive capabilities. Notwithstanding our strength in the field of ICT, such capabilities, skill sets and the much needed synergy amongst all concerned in the nation would take a long time; assuming that we have and can display political will and resolve. A few initial steps have been taken but we have a long way to go. The nation urgently needs to develop a credible cyber deterrence through promulgation of a doctrine, development of the skill sets, a dedicated organisation and demonstrable cyber defensive and offensive capabilities.

## Endnotes

1. The rise of Cyber Weapons and Related Impact on cyberspace. Resources. infosecinstitute.com dated Oct 5, 2012

2. Stuxnet: A malware said to have been created under a special programme, "Operation Olympic Games" jointly by the USA and Israel and used to damage the centrifuges at the Natanz nuclear facility of Iran.

3. What is a cyber weapon ? Paolo Passeri, April 22, 2012. Hackmageddon.com

4. Ibid, endnote 1.

5. Ibid.

6. Generations of Cyber Weapons. Selil.com/archives/3152, July 10, 2012

7. Estonia : In the spring of 2007, a cyber attack on Estonia blocked websites and paralysed the country's entire internet infrastructure.

8. Georgia: In August 2008, Russian planners tightly integrated cyber operations with the kinetic, diplomatic and strategic communication operations and achieved cyber disruption at the moments they needed those disruptions to

occur.

9.  Kaspersky Security Bulletin 2012 : Cyber Weapons

10. Wiper: One of the world's most powerful data snatching cyber espionage virus targeting computers in Iran, Israel and middle eastern countries. The worm has allegedly been used in state sponsored espionage.

11. Gauss : A new cyber surveillance virus found in West Asia that can spy on banking transactions and steal log-in information from social net working sites, e-mails and instant messages. Discovered as part of United Nation backed effort to reduce the global impact of cyber weapons.

12. Duqu: A new virus which can leak data from any computer it infects – particularly leaking data from power plants, oil refineries and pipe lines. Data acquired by this virus is used for creating new cyber weapons.

13. miniFlame: A small and highly flexible malicious programme designed to steal data and control infected systems during targeted cyber espionage operations. It is a high precision surgical attack targeted cyber weapon used as the second wave of cyber attack.

14. Cyber weapons: Leveling the International Playing Field. Ross M Rustici

15. Ibid, endnote 3.

16. Ibid.

17. Global Cyber Deterrence- East West Institute. www.ewi.info/system/files/ CyberDeterrence.pdf Cyber Weapons by P Paganini, April 3, 2012, Hacker News

18. Ibid.

---

†   Lieutenant General Davinder Kumar, PVSM, VSM and Bar (Retd) was commissioned in the Corps of Signals in December 1965 and retired as Signal Officer-in-Chief in September 2006. Post retirement he has worked in the Corporate Sector, contributed extensively in professional journals and has been a member of National IT Task Force and a number of Committees.

# Cyber Space, Outer Space and Information Space as the Non-Linear Strategic Frontiers[*]

## Lieutenant General Davinder Kumar, PVSM, VSM & Bar (Retd)

It is widely believed that 21 century warfare will be fought largely in the virtual domain, thus catapulting Information Operations, conducted in cyber space, to a core national capability along with air, land, maritime, space and special operations. In the digital battle field of twenty first century, Technology and Information are the new normal and Cyber space, Outer Space and Information space the new domains.

Information has been recognized as a strategic resource which must be effectively managed to maintain a competitive and evolutionary advantage. Because of its critical role in reducing uncertainty, structuring complexity, and generating greater situational awareness, any action taken in the information domain can leverage tremendous effects in both the physical and cognitive domains. Denial and manipulation of information are important instruments affecting the hearts and minds of a nation's population, the governance and the availability of critical systems and services and so on. Consequently, information has become critical to both the national sovereignty and military success. It also extends the range of new options for a planner or decision maker. The cyber space and outer space largely reside in the Information space, also called "Infosphere" and together they represent the new strategic frontiers at the heart of a nation's comprehensive combat power.

---
[*]  This article was first published in the *USI Strategic Year Book 2017*.

# Cyber Space

Cyberspace is the proverbial ether within and through which electromagnetic radiation is propagated in connection with the operation and control of mechanical and electronic transmission systems. *A precise definition would be that [1]Cyber Space is "A global domain, within the information environment, whose distinctive and unique character is framed by the use of electronics and electromagnetic spectrum to create, store, modify, exchange and exploit information via interdependent and inter-connected networks using information communication technologies".*

## Comments

- Cyberspace is an operating domain just like the atmosphere and space, and it embraces all systems that incorporate software as a key element.

- It is an operational medium through which "strategic Influence" is conducted.

- The fundamental condition of cyber space is the blending of electronics, information and electro-magnetic energy.

- It is a medium, in which information can be created and acted on at anytime, anywhere, and by essentially anyone.

- It is qualitatively different from the land, sea, air, and space domains, yet it both overlaps and continuously operates within all four.

- It also is the only domain in which all instruments of national power (diplomatic, informational, military, and economic) can be concurrently exercised through the manipulation of data and gateways.

- Cyberspace can be thought of as a "digital commons" analogous to the more familiar maritime, aerial, and exo-atmospheric commons.

- Like the other three commons, continued uninhibited access to cyber space can never be taken for granted as a natural and assured right.

- Unlike, other domains, cyber space is a domain in which the classic constraints of distance, space, time, and investment are reduced, in some cases dramatically, both for ourselves and for potential adversaries.

- In the global economy of 21st century, cyber space is perhaps the single most important factor linking all the players together, boosting productivity, opening new markets, and enabling management structures that are flatter yet with a far more extensive reach.

## Cyber Space-a Non-Linear Strategic Frontier

Cyberspace has arguably become a *Nation's center of gravity* not just for military operations, but for all aspects of national activity, to include economic, financial, diplomatic, and other transactions. It has been explicitly recognized as an operating arena at par with the atmosphere and space and begun to be systematically explored as a medium of combat in and of itself.

It provides the capability to launch organized attacks on critical infrastructure and other targets of interest from a distance, on a wide variety of "fronts," and on a global scale, at the speed of light. It is an economical tool for "*power projection*" and has altered the concept of sovereignty. It is the principal domain in which a Nation exercises its command, control, communications, and ISR (Intelligence Surveillance Reconnaissance) capabilities, that enable global mobility and rapid long-range strike.

Cyberspace is transforming into a vast, complex universe, and we need new tools to understand it. [2]*It is dangerously deceptive because it linearises a dynamic process that is highly non-linear.* There are several non-linear, dynamic processes in play that are driving the evolution of cyberspace as a complex system. And like many complex systems, it is inherently unpredictable. [3]The Web is shifting power in ways that we could never have imagined'

[4]*Cyberspace is 'reinventing warfare'. For the major armies of the world, formed by the conventions of the Industrial Age, twenty-first century conflict* seems unfathomably complex and ambiguous, because of the impact of technology and information, which are not only redefining "Warfare", but also the manner of its execution.

[5]Cyberspace increases transparency. It makes it harder to keep things secret; and this, in turn, makes it harder to govern. Without doubt, cyberspace enables movements to mobilize rapidly in forms that are highly fluid and hard to defeat. The 'landscape of international relations' has been changed

irrevocably by the way in which information now flows more freely into the public sphere.

## Distinguishing Characteristics of Cyber Space

- Control of cyberspace is a sine qua non for operating effectively in the air, maritime and outer space domains.

- Denial of unimpeded access to the electromagnetic spectrum through hostile actions, would render satellite-aided munitions useless, disrupt command and control systems, and the ensuing effects could be paralyzing. Accordingly, *cyberspace has become an emergent theater of operations that will almost surely be contested in any future conflict.* Successful exploitation of this domain through network warfare operations, can allow an opponent to dominate or hold at risk any or all of the global commons. One reason for the imminent and broad based nature of the cyberspace challenge is the low buy-in cost compared to the vastly more complex and expensive systems of air and space warfare. Consequently, smaller adversaries are able to cause *"catastrophic cascading effects"* through asymmetric operations against stronger and developed nations.

- Cyberspace warfare is offense dominant wherein organized attacks on critical nodes of the nation's infrastructure can be launched by any party. Such attacks can be conducted both instantaneously and from a safe haven anywhere in the world, with every possibility of achieving high impact and a low likelihood of attribution and, accordingly, of timely and effective retribution.

- Vulnerabilities in cyberspace are open to the entire world and are accessible to anyone with the wherewithal and determination to exploit them. *The challenge is to find them and that is one of the prime tasks of cyber espionage.*

*Any loss of cyberspace dominance would negate gains, to a large extent, in air and space in virtually an instant.* Many nations, governments, militaries, private sector, R&D establishments and Academia across the world are, accordingly, involved in developing technologies for offensive cyber operations. *All of this has rendered offensive cyberspace operations an attractive asymmetric option for state, non-state actors and introduced a new but potent threat in the form of "lone wolves".* The most sophisticated threat may come from China, which unquestionably is already a peer competitor to both the USA and Russia, with ample financial resources and technological expertise.

State-sponsored cyberspace intrusions are now an established fact and have been largely accepted. These have become a major source for espionage by way of probing attacks, technical and industrial intelligence. In light of its relative newness as a recognized and well-understood medium of combat, detailed and validated concepts of operations for offensive and defensive *counter–cyber warfare* and *cyberspace interdiction* are, in all probability, been formulated by different entities and integrated as part of the comprehensive national power. *The bottom line is the development of full spectrum cyber power with a view to, "dominating the electromagnetic spectrum—from wired and unwired networks to radio waves, microwaves, infrared, x-rays, and directed energy."* Such activity would require production of appropriate cyber weapons which can be both surgical and mass-based in their intended effects.

## Outer Space

[6]In the information age, satellites have become a core element of modern societies and are largely responsible for bringing the nations and individuals together. While satellite-based communications and navigation systems help to improve traffic safety, disaster response, or weather forecasts; satellite based systems help in education, health, earth resource management and so on. Global climate change and the concomitant increase of water conflicts and energy crises will further enhance the importance of satellites *as means of information procurement and disaster response.* The advancement of the information society will also create new vulnerabilities. The more societies depend on satellites, the more important it will be to protect them as critical infrastructures. For strategic reasons, the vulnerability of space based systems used for collecting and relaying security-relevant information is increasing. *Today, there are two general developments jeopardizing the safe and peaceful use of space: First of all, increasing space pollution, and secondly, the re-emergence of arms dynamics in space.*

*In a space system, most of the technologies have a dual-use character and civilian satellites are increasingly being used for military purposes.* [7]For modern armed forces, Satellites have become indispensable, especially considering the irresistible advance of network centric warfare since the war in Afghanistan from 2001 onwards. This involves the integration of information from various military platforms, such as tanks, vessels, or aircraft, into a jointly used information network that optimises decision-making processes and navigation of forces. *Satellites thus serve as force multipliersin the present day military operations.*

Space based systems and satellites will continue to increase in importance. There is a new sensitivity and urgency emerging about the security of outer space and related techno-military superiority in space. [7]Military blueprints by major space-faring powers now encapsulate concepts of *'space support'* and *'force enhancement'* which point to a central role of space assets in facilitating military operations while notions of [8]*'space control', "space power", "space situational awareness"* and *'force application'* suggest that not only the space has been recognized as the new strategic frontier but its weaponisation is likely in near future. The majority view indicates that space may, in the near future, be a theatre of military operations. [9]Counter-space capabilities such as missile defence, anti-satellite capabilities and a new class of Directed Energy Weapons (DEWs), thus, assume critical importance for defence and security perceptions. Space faring nations across the world are busy in developing technologies and capabilities for physical and electronic hardening; anti-jam technologies, satellite maneuverability, redundancy at system and sub-system levels; quick launch facilities, mini, micro and Nano satellites both for restoration of facilities and as killer satellites.

## Cyber Space and Outer Space

Cyber space is relatively a new strategic frontier as compared to outer space. It can, therefore, learn and adapt "good practices" evolved in space capabilities particularly in the realm of joint warfare. The logical approach in this exercise would be towards identifying gaps, shortfalls, and redundancies in existing offensive and defensive capabilities. Also, parallels between space and cyberspace, as domains of offensive and defensive activity, can be used for developing and validating cyberspace concepts of operations, tactics, techniques and procedures. [10]For example:-

- Both domains, at least today, are principally about collecting and transmitting information.

- Both play pivotal roles in enabling and facilitating lethal combat operations by other force elements.

- Both, have more to do with the pursuit of functional effects,than with the physical destruction of enemy equities, even though, both can materially aid in the accomplishment of the latter.

- In both domains, operations are conducted remotely by warfighters sitting before consoles and keyboards, not only outside the medium itself, but also in almost every case out of harm's way.

- Both domains are global rather than regional in their breadth of coverage and operational impact.

- Both domains overlap—for example, the jamming of a GPS (*Global Positioning System*) signal to a satellite-aided munition guiding to a target is both a counter space and a cyberwar operation insofar as the desired effect is sought simultaneously in both combat arenas.

- Space provides an overarching capability to view globally and attack with precision from the orbital perspective. Cyberspace provides the capability to conduct combat on a global scale simultaneously on a virtually infinite number of fronts.

## Information Space

[11]The free flow of information within and between nation states is essential to business, international relations, social cohesion and to a military force's ability to fight. We live in a highly connected world, but it doesn't take much to tip over into instability or even chaos. The experience of recent conflicts, globally, one can easily conclude that the information is one of the most important tools in the hands of the military decision-makers.[12]Information is a strategic resource vital to national security. Military operations today, are very intensely dependent on information and information systems both for effect based operations and integration of different components of national comprehensive power. Information is equally important in the planning and execution of operations at operational and tactical levels. [13]Today's way of fighting wars has changed, significantly influenced by the technological revolution in gathering, storing, analysing and dissemination of information. The speed, accuracy and timeliness of information are at the core of the concept of information operations. It is imperative that military and civilian leaders understand the true value of "*information management*" and "*information space*" and the absolute necessity of achieving *information*

*superiority* in a conflict.

"*Information Space* is a sum of individuals, organizations and systems that can collect, process, distribute or act on information". Information management is about quick processing of the raw data and presentation of useful and corroborated information at correct time, in relevant format and at the right place. People and automated systems observe, orient, decide and act on the information, in the information space which is the main area of decision making. [14]Lead in "*competitive decision cycle*" is the essence of Information Dominance. Superiority in competitive decision cycles requires one side to understand what is happening and act faster than the other. Simply stated, it is the ability to "*operate inside an adversary's Observation-Orientation-Decision-Action (OODA) loops or get inside his mind-time-space to penetrate his moral-mental-physical being in order to isolate him from his allies, pull him apart and destroy his will to resist*". This is precisely the principle underlying the Revolution in Military Affairs.

Information superiority is essential to achieve almost all joint combat capabilities of today's modern battlefield. By nature it is a relative concept and is transient. It is created and maintained by conducting information operations. It provides a competitive advantage only when it is effectively translated into superior knowledge to achieve "*decision superiority*", better decisions, adopted and implemented faster than the opponent can act, or in non-combat situation, decisions taken at a pace that allows forces to shape the situation or react to changes and achieve their mission.

*Information Space exists within each of the four domains: sea, land, air and space and affects the ability to perform military operations in three dimensions - physical, information and cognitive.* Accordingly, Information/Cyber space has been recognized as the Fifth domain of warfare. [15]Seeing the exponential penetration of Information and Communication Technologies, and Media and their capabilities to influence and penetrate human mind, the day is not far when "*Human Mind*" will be the sixth and perhaps the most potent domain of warfare.

War in the sixth domain would be about controlling the human mind, either by shaping emotional and cognitive responses, or by outright exploitation of man-machine technology. It is, in a sense, *coercive persuasion through internal and external stimuli.*

Information operations conducted in the cyber space and the instant global reach provided by the outer space would be at the heart of this *"Neuro Warfare"*. It is high time that we deploy adequate resources both in development of and protection from such operations and enhance our information warfare capabilities and undertake techno-military orientation of our space programmes. *Space, Cyber and Infosphere are the emerging strategic frontiers in the 21ˢᵗ century security scenario.* Development of capabilities, to conduct defensive and offensive operations to secure them are an absolute necessity both for human and national security. India needs to work on these in a mission mode and integrate these capabilities to enhance her comprehensive national power relevant to the likely threats and warfare in the present century.

## Endnotes

1. Cyber Security-a National Strategic Imperative by Lieutenant General Davinder Kumar.

2. The New Final Frontier; Asia and the Pacific Policy Studies https://asiaandthepacificpolicystudies.crawford.anu.edu.au/news.../

3. Ibid

4. Cyberpower in Strategic Affairs: Neither Unthinkable nor Blessed DAVID BETZ Department of War Studies, King's College London

5. Ibid

6. Space: The New Frontier of Security Policy - CSS www.css.ethz.ch/publications/pdfs/CSSAnalyse171-EN.pdf

7. FROM STAR WARS TO SPACE WARS—THE NEXT STRATEGIC FRONTIER:PARADIGMS TO ANCHOR SPACE SECURITY Jackson NyamuyaMaogoto* and Steven Freeland**

8. Cyber, Outer Space and Disruptive Technologies, by Lieutenant General Davinder Kumar

9. Fusion of Technology, Cyber and Space Elements of Hybrid Warfare, by Lieutenant General Davinder Kumar

10. Ibid

11. What is information warfare? | World Economic Forum https://www.weforum.org/agenda/2015/12/what-is-information-warfare/

12. INFORMATION AS A STRATEGIC RESOURCE CRITICAL TO eprints. ugd.edu.mk/.../INFORMATION%20AS%20A%20STRATEGIC%20 RESOUR...1.by M Bogdanoski - 2015

13. Ibid

14. INFORMATION WARFARE - Developing a Conceptual Framework

15. Is War in the Sixth Domain the End of Clausewitz? by Chloe Diggins & Clint Arizmendi.

†

---

† *Lieutenant General Davinder Kumar, PVSM, VSM & Bar,* retired as the Signal Officer-in-Chief of the Indian Army in September, 2006, after rendering 41 years of distinguished service. He was the CEO & Managing Director of Tata Advanced Systems Ltd from September, 2008 till September 2011. He has worked with Indian Space Research Organisation (ISRO), Oil India, and the Planning Commission and has been on the Board of Public and Private sector companies. He has over 400 papers to his credit and has spoken on various national and foreign fora forums